FIRE ON THE HORIZON

FIRE ON THE HORIZON

THE UNTOLD STORY OF THE GULF OIL DISASTER

John Konrad and
Tom Shroder

HARPER
An Imprint of HarperCollins*Publishers*
www.harpercollins.com

HarperCollins books may be purchased for educational, business, or sales promotional use. For information, please write: Special Markets Department, HarperCollins Publishers, 10 East 53rd Street, New York, NY 10022.

FIRST EDITION

Library of Congress Cataloging-in-Publication Data

Konrad, John.
Fire on the horizon : the untold story of the Gulf oil disaster / by John Konrad and Tom Shroder.—1st ed.
p. cm.
ISBN: 978-0-06-206300-7
1. BP Deepwater Horizon Explosion and Oil Spill, 2010. 2. Underwater explosions—Mexico, Gulf of. 3. Offshore oil well drilling—Mexico, Gulf of. 4. Petroleum industry and trade—Accidents—Mexico, Gulf of. I. Shroder, Tom. II. Title.
HD7269.P42M615 2011
363.119622338190916364—dc22 2010051657

11 12 13 14 15 OV/RRD 10 9 8 7 6 5 4 3 2 1

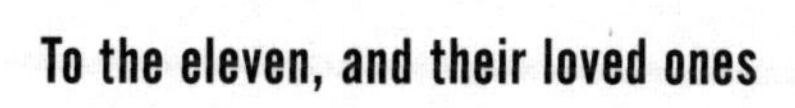
To the eleven, and their loved ones

CONTENTS

AUTHOR'S NOTE: A PERILOUS CROSSING

In the vast southern ocean, below the continental tips of South America and Africa and above the ice of Antarctica, storms take complete loops around the earth with no land to obstruct or diminish their force. Occasionally one of those storms veers north to sandwich ships between high winds and land as they round the Cape of Good Hope. On April 19, 2010, my ship broke through just such a veering storm that had brought with it sheets of rain and gusts of hurricane-force wind. Fortunately, I was sailing on the drillship Deep Ocean Ascension, the latest, most expensive and technologically advanced of BP's fleet of exploratory drillships, built in a post-Katrina world to handle the extremes of nature—up to fifty-foot waves and winds of 115 miles per hour—with 54,000 horses of available power to propel 105,000 tons of equipment and steel.

I had been with the ship since she was a collection of scattered parts in a Korean shipyard; I was serving as acting captain to a vessel that had yet to float. The only navigating I was responsible for at that point was maneuvering around construction delays and bottlenecks to ensure that she, and particularly her safety systems, were built to specification. Any foreseeable emergency that would visit the ship had been considered before construction had even begun; a new drilling rig is, if anything, a rig made wiser by the disasters of its predecessors. The prospect of fire is anticipated everywhere: the foam dispensers mounted above the accommodations, the deluge sprinklers, lockers filled with firefighting gear, and the rows upon rows of tall, thick canisters above the engine room that, through countless pipes within the rig, could blanket machinery spaces with more than ten thousand pounds of fire-snuffing CO_2. No expense was spared for the safety of this $750 million vessel.

Yet danger remained. I had continued my training since graduating from SUNY Maritime College in 2000 and had spent nearly ten years advancing my licenses to the highest level–master unlimited—but, as the son of a fireman whose company, Rescue 3 of the Bronx, lost its crew on 9/11, I was aware that it's not the dangers you anticipate but rather the unforeseen failures that most often give birth to a catastrophe. And at sea and in the Gulf, when disaster strikes, there are no 911 services to call.

This fact wasn't lost on my wife, Cindy. A mariner herself, she had navigated a large containership stocked with ammunition and supplies through the combat zones of the Persian Gulf. She realized that the danger of rounding the Cape of Good Hope was small compared with navigating waters mined with explosives, but life had changed with the birth of our children, Jack and Eleanor. I'd been spending longer stretches at our California home, and I'd launched a blog and networking website for mariners that had diminished the financial imperative for me to spend long stretches at sea.

And yet here I was, again.

If anyone had told me ten years earlier that I would come to love life offshore, I'd have laughed. But I had discovered a reality that few understand or appreciate: the offshore oil field is a magic place where people pit technology against nature to accomplish impossible tasks. It's a place that more often than not rewards hard work, intelligence, and determination; where degrees and résumés don't matter; where even a high school diploma is not necessary to lead divisions of men and women.

So in 2009 I accepted an assignment with Pride International on the as yet unfinished Deep Ocean Ascension. Now, months later, in mid–April 2010, we were sailing through this storm near the Cape of Good Hope.

The weather cleared as the storm passed, and the next day, April 20, we awoke to clear skies and smooth seas. It was 5:30 in the morning when we had our first meeting, but back in our Houston headquarters the time was 9:30 p.m. An hour passed and I was in my office working on paperwork when Cameron Whitten, the ship's second officer, scrambled up the stairs with a confused look on his face. He had just been to the crew meeting where one of the guys, calling home after a late shift, got word of a blowout on a rig, owned by the world's largest offshore drilling company, Transocean Ltd., named the Deepwater Horizon.

I turned on my computer in disbelief. It had to be wrong. Google, CNN, and the rest of the online world made no mention of the event. Then I turned to my blog, gCaptain, and I saw it. A longtime reader, Captain Thad Fendley, was laying anchors for another Transocean rig just ten miles away from the Horizon, snapping pictures and posting the first public images of the rig in flames. I stared at the three-hundred-foot-high fireball that seemed to be consuming the entire Horizon. My heart plunged in my chest; my eyes started to well. My anguish was reflected all around me. A half-dozen among the Deepwater Ascension crew had previously worked on the Horizon. Countless others had friends, neighbors or family aboard. One's brother had been a member of the crew working around the derrick, where the explosion had ignited that night.

The Horizon wasn't just any rig to me. I had worked for Transocean for seven years, and I had spent nearly ten years (on and off) on ships contracted to BP, the company for which the Horizon was drilling.

Faces of those I knew on the Horizon flashed painfully in my mind. Mark Hay, the subsea engineer on my first rig, the Discoverer 534, had been generous sharing his knowledge with a green hand. Mike Mayfield had reported directly to me on the rig Discoverer Spirit, and had been a willing teacher to a boss two decades younger than himself. Curt Kuchta, a friend with whom I had risen through the Transocean ranks, was the Horizon's captain.

And Dave Young. Dave Young, the Horizon's chief mate, was one of my closest friends. We'd met in 1996 as second-year students at SUNY Maritime. We'd sailed the world together in the academy's training ship. We'd stayed close through the years as we both married and started families. In fact, I was the reason Dave was on the Horizon—for years I'd invited him to come to work in the oil field. I was still with Transocean in 2007 when he finally agreed to apply, and I'd pulled whatever strings I had to help get him in the door.

I kept looking at the picture of the all-consuming fire, still raging, I knew, five thousand miles away to the west and north.

It was late at night back home. Did I call Dave's wife, Alyssa? My God! They'd just had their third child a few weeks before. I began waking friends with Transocean in Houston, and I followed the updates flowing in from Thad Fendley and the other boat captains who were participating in the rescue efforts and updating everyone via gCaptain, which by now had gotten the attention of many of the Horizon crew's families, who were posting frantic pleas for information.

But for long hours, nobody knew the fate of the crew. No one had heard from those who were aboard the Horizon. No one knew if Dave was still alive.

I watched a crewmate dial the numbers for the onboard phones on the Horizon. They just rang and rang and rang.

That night, twelve hours after we first heard the news, the captain passed word that all crew had been rescued. Excitement and relief spread across the ship as we approached Cape Town, South Africa. But soon our elation turned back to grief with the word that eleven men were missing; before long, that turned to "hadn't made it."

I was relieved so many had been rescued, but I couldn't get those who would never come home to waiting friends and family out of my mind.

As the ongoing ecological catastrophe caused by the burning and then the sinking of the Horizon came to dominate the news, the focus of the public quickly shifted to the marshes, beaches, and waterways threatened by the spreading black cloud of oil, and to the impotence and frustration of multiple failures to stop the flow from the ruptured well a mile beneath the ocean's surface.

Classmates of Dave's and mine from SUNY Maritime were intimately involved in the work to stem the blowout. Richard Robson and his crew on the Development Driller III worked around the clock—swarmed by media, BP officials, engineers, and government officials—to drill the relief well that would stop the flow of oil. Simultaneously, another close friend and classmate, Matt Michalski, captain of the Development Driller II, positioned his rig next to Rich and began drilling the second relief well. The Horizon blowout had riveted a nation, but in a far more personal way it had fully consumed our close-knit group and the wider industry of deepwater drilling to which we belonged.

In the massive coverage that followed, in the finger-pointing and eye-crossing dissection of technical blame, I saw only jagged fragments of the full reality of the tragedy. I came to believe that what happened on the Deepwater Horizon, over block 252 of the subsea geological formation known as the Mississippi Canyon of the Gulf of Mexico, could never be completely understood without placing it in the full context of the powerful, in many ways inspiring, but also intrinsically flawed and little-understood culture of offshore drilling.

For all my years in the industry, up until April 20, 2010, the world of deepwater drilling had been an obscurity, one I would have difficulty explaining even to family and friends. Now I knew that was no longer acceptable. With my personal experience, my connections to the Horizon, and the many links I'd forged within the industry through my work with gCaptain, I felt uniquely placed to attempt to tell the story, the as yet untold story, of the Deepwater Horizon from the perspective of the people who lived and died at the leading edge of the struggle to decrease America's dependence on foreign oil. Whatever your politics, or your feelings about the use of fossil fuels, the fact remains that the world continues to need and demand immense quantities of oil, and in the end, we all rely on these crews of strong, skilled, and determined men and women to bring it home. It's our obligation to also understand the risks and the pitfalls they face in doing so.

I blog daily about the ships and rigs whose stories fascinate me, but I am not a professional writer or a journalist. That is why I have partnered with my co-author, Tom Shroder, a writer and editor with thirty-five years experience at some of the nation's most respected newspapers, most recently *The Washington Post.*

We did not want this book to be a political argument, or even a judgment on ultimate responsibility for the disaster. The investigations and legal cases that will eventually make those determinations are ongoing, and probably will be for years.

This is not the story of a rig, technology, the environment, corporate policy, or government oversight, but it concerns each. This is the story of men and women. Good men and women facing unprecedented technological challenges under unparalleled economic pressures. It is the story of a way of life, and a devastating tragedy that those challenges and that way of life made only too inevitable.

Nonetheless, I have a personal history that might prompt some readers to look for an element of bias in this book. While working for Transocean in 2008, I found myself balancing the needs of my crew with the rapid growth of a company working through a large merger and expanding to meet the needs of the energy bubble. I got into a dispute with company management over safety procedures. As a result of the dispute, I was placed on leave, and then finally administratively discharged without notice of fault. As I have mentioned, I went on to other jobs in the industry with companies including BP and to develop gCaptain. Despite that dispute, there are still many things about Transocean I respect and admire, and many, many people who I worked with there for whom I have deep affection.

I can attest that Tom and I have desired only to reflect the Deepwater Horizon's reality as truly as we can. To that end we have based our account on interviews with firsthand sources, thousands of pages of sworn testimony, official reports, and public documents. I ask that you read this book, and judge for yourselves if we have succeeded.

—John Konrad

PROLOGUE: THE END

2315 Hours, April 20, 2010

Doug Brown, chief mechanic of the Deepwater Horizon, had survived the direct impact of two explosions, guided an injured man to the bridge around piles of debris, and made a harrowing exit from a burning oil rig. But only as his lifeboat pulled away did he realize something was wrong with his left leg. It began to ache, and then the pain blossomed into a fiery torment. The too-recent memory of hiking all over the rig, skirting flames, climbing up and down stairs, seemed like an impossibility. He was a sick kind of dizzy and his head throbbed. The hard bench had become a torture device prodding his lower back into a clenching knot that had grown tighter with each indifferent drop and roll of the lifeboat. He'd find out the physical toll soon enough: a fractured fibula, torn knee tendons, nerve damage, loss of feeling in a softball-sized chunk of his calf, lumbar strain, and a concussion. The psychological and emotional cost would be more difficult to calculate.

After he'd been lifted to safety aboard the rescue boat, there was nothing to do but sit on deck, wait for a medevac helicopter, and watch flames shoot through the top of the Horizon's derrick a half mile away. It seemed a trick of perception, one of those out-of-body experiences he'd read about. In his mind he was still on the rig, still feeling the heat and the fear. But at the same time he was watching it burn across a dark void. His inexpressible sadness for the Horizon and what it meant—a home and way of life—merged with the pain of names left blank on a muster list. He'd known all eleven of the missing, counted himself close friends with five. There wasn't anything he could do about any of it, except sit and watch the flames consume the Horizon. There was nothing to do now but witness her death, just as he had witnessed her birth.

CHAPTER ONE

THE BEGINNING

December 2000

Ulsan, South Korea

Half dead from lack of sleep, Doug Brown was staring out the bus window at the gates of the Hyundai Heavy Industries shipyard in Ulsan, South Korea, five thousand miles from home, when a raging flood of scooters and mopeds burst through the winter morning's fierce grip and woke him up for good. As the violent buzz bore into his eardrums, Doug recoiled at what his American colleagues called the "Hyundai 500," a name that didn't sound so sinister back in the States. Some of the scooters were doubled up and stacked so high with bundles and packages that the slightest bump would surely have knocked one into the next and sent the whole mass crashing like dominoes. It seemed to Doug that all ten thousand workers at the world's largest shipyard were arriving at the same instant, moments before the 0900 start time.

What he could see of the yard through the window looked like an industrial version of Disney World: walkways lined with flowers, buildings seemingly wet with fresh coats of paint, impeccably dressed security guards with helmets and badges buffed to a pristine, reflective shine. Uniformed signalmen positioned on platforms in the center of the road moved robotically as the chaos swarmed below. The whole crazy scene was framed by five-story-high gantry cranes straddling immense drydocks at the water's edge, and seagoing vessels laid section by section on a blacktop the size of a Walmart parking lot, as if playful giants had abandoned their toys there.

Rising from the drydocks and floating at mooring cables were more finished or nearly completed ships than at any other shipyard in the world. The vessels were of astonishing variety and power—liquefied natural gas carriers crammed with cryogenics and studded with insulated domes, each the size of a small basketball arena; supertankers capable of carrying two million barrels of oil; car carriers with their tall boxlike hulls and internal maze of ramps capable of holding thousands of vehicles; and countless container-ships without a single container on deck.

If the scooter stampede hadn't so thoroughly woken him, Doug might have believed he was still dreaming. He'd only been working for the offshore oil company R&B Falcon for two years when his managers pulled him aside and offered him a prestige assignment: chief mechanic of their newest, most technologically advanced rig, still being built in Korea. That was barely a month ago, and now he was in this fantastical place halfway around the world, about to see his new rig—his new home—for the first time.

Doug had grown up in a middle-class family in California. He was a round-faced, round-figured man of average height with kind blue eyes and a laid-back temperament in that open, Pacific Coast way. Nothing about him said striver, yet here he was making a rapid climb in one of the fastest-growing and most profitable industries in the world. He didn't even have a college degree, but what he did have was a natural ability to work with machines, and experience.

He'd started out working in construction and landscaping. When the lawn tractors broke down, he was the one who got them running again. He had an intuition about engines, a feeling and liking for them. For most people an engine is an unorganized cacophony of sounds, but to Doug the fast beats of compression and ignition formed an intelligible rhythm. He could tell when a piston misfired by the sound of a missed beat. He had a talent, but no formal training, something he thought he could rectify when he decided to join the army in 1989, at the age of twenty-nine. He started out learning to maintain helicopter engines, then migrated to an oxymoronic job: working on army marine engines during the Persian Gulf War—landing craft, tugboats, and barges used to ferry troops and equipment across rivers and between ports. The closest he and his heart rate came to combat were the poison gas alarms that screamed across his base in Kuwait, signaling the launch of Iraqi Scud missiles in his general direction.

He learned a little about working on the edge of danger, but even more about the complex machines that powered seagoing vessels. Engines were engines—gas had to be delivered from a tank to a chamber where it would mix with air, compress, explode, and drive the movement of a shaft—but ship engines were larger, with more moving parts. He also had to learn the mechanics of auxiliary systems needed on a self-contained vessel at sea: water desalination, ventilation and cooling, hydraulics, even the processing of sewage. While managing these systems lacked the glory of combat infantry, the skill set he'd developed paid off when his army career approached an end in 1998 and he posted his résumé online.

The offers all came from offshore oil companies. He accepted the one from R&B Falcon. The company was not the largest or the most profitable. If Doug had had a background in finance its balance sheet might have caused him to reconsider. But R&B was pushing into the new frontier of deepwater drilling with an aggressive program of building new rigs, each among the most advanced in the world, and, most importantly to Doug, they offered the highest pay. He wasn't married, so the schedule of twenty-one days on and twenty-one days off didn't present any family problems, and the rig's atmosphere—the smell of burning diesel and the constant grind of heavy equipment, the hierarchical command structure—all reminded him of the military.

He signed on as a mechanic on a rig in the Gulf of Mexico, was promoted quickly to chief mechanic, and had barely settled in to his new supervisory responsibilities when he was swept up in this Korean vortex.

As quickly as the storm of scooters appeared, it passed. The flow slowed to a trickle, and the bus lurched forward, carrying Doug and his future rig mates on a clear path to the docks. As they approached their destination, Doug cocked his head flat against the cold window to get a glimpse of the derrick towering above the Deepwater Horizon.

The Horizon had begun life as simple sheets of thick steel, lifted from trucks by powerful magnets and placed on a gigantic conveyor belt. First stop was an industrial-strength version of Goldfinger's powerful laser cutter, which sliced the metal into the proper shape. From there the cut steel rolled into the bending room. No machine had yet been invented for the task of shaping such thick steel, so the yard relied on an age-old manipulation by fire and water. The steel was laid down on prefabricated plywood molds, then, with a fire torch in one hand and a garden hose in the other, the master bender, sitting directly on the steel on a wooden stool, heated the metal till it began to wilt over the plywood frame. Once it fit the form, he'd douse it with water, cooling and immediately hardening the steel. Now cut and shaped, the steel sheets were welded to form tanks, rooms, bulwarks, derrick sections, or any one of the hundred other modules of a rig; then every inch was painted. The finished sections were carted outside to the asphalt assembly lot. There they were outfitted with pipes, electrical cables, gauges, hoses, and the myriad other small components, then stacked and welded together to form massive megablocks, the biggest ones twelve stories high and one to two hundred feet square.

The job of moving these behemoths around the yard fell to oddly small devices called transporters. Just a few feet tall, the transporters had a flat deck, beneath which were 144 small wheels connected to powerful hydraulic jacks. The short profile of the transporters allowed them to crawl underneath the megablocks, where they were linked by a wireless signal from a computer. They moved in unison into position, then the hydraulics engaged, elevating the transporters and lifting the entire megablock off the jacks. Traffic stopped, the roads cleared, and the entire immense assemblage rolled down to the dock. In a short few weeks, huge cranes on barges lifted each megablock into its precise position. Now the growing skeleton was swarmed by workers donning masks, starting generators, lifting spray guns, each performing a single task that inched the rig closer to completion. At first the welders dominated. Men and women wearing leather smocks and face masks covered every corner of the structure, maneuvering across scaffolding and through holes in the bulkhead, which they would eventually weld shut. Painters followed behind the welders, laying coats of protection over the steel. As the rig got closer to completion the ratio of welders to painters tipped and the metallic smell of burnt steel was replaced with the noxious fumes of drying paint. Then came electricians, pipefitters, electronics techs, quality control engineers, and a myriad of other specialized workers who added layers of complexity until the rig was ready for its final assembly.

This was where Doug came in, about three months before the rig was scheduled to be launched. He and his busload of future rig mates, each of them handpicked from other company assets, arrived as the multi-tiered deck was hoisted one hundred feet in the air by some of the world's largest cranes, then settled carefully atop the four columns rising from the pontoons.

To the uninitiated, the Deepwater Horizon was far from beautiful. None other than John Steinbeck once described one of the Horizon's early ancestors as having "the sleek race lines of an outhouse standing on a garbage scow," and nearly a half century of technological advances hadn't changed that.

In place of a single hydrodynamic hull, the rig would float on its two long, narrow pontoons, each twice the size of a 727's fuselage. The pontoons were lined with computer-controlled ballast tanks that could be flooded or emptied in precise increments that finely adjusted the rig's trim, and the main deck's elevation above the ocean surface.

When the rig moved from one spot to another on its own, powered by eight 7,000-horsepower electric thrusters mounted beneath the pontoons, the ballast would be adjusted so the pontoons rode thirty feet below the surface, submarine-like, and the deck towered more than a hundred feet above the water. When the rig moved over a well and prepared to drill, the pontoons would take on more water and sink to a depth of seventy-six feet, creating a more stable base and lowering the deck to just sixty feet above the sea.

The four massive support columns looked like highway trestles, or the limbs of a brontosaurus, supporting what appeared to be a crowded construction site in the early stages of the creation of a small skyscraper. The Eiffel Tower–like steel structure in the middle of the rig was an oil derrick, rising to 260 feet above the deck from a forty-eight-foot-square base. Two built-in cranes to either side of the tower were designed to transfer steel pipe and other heavy equipment used in well drilling from supply boats to what was known as the rig floor—a platform surrounding the derrick one level above the main deck. Clinging beneath the main deck was the "accommodation block," a two-deck honeycomb of steel that was basically a combination hotel and office building. In addition to crew cabins, conference rooms, and offices, there was a kitchen, a dining hall, a movie theater, a gymnasium, an infirmary, and a lounge—everything needed to house, feed, care for, and entertain the full complement of 126 oil workers, sailors, and managers.

Some of those top managers, beginning with the offshore installation manager (always called the OIM), who was the top on-rig authority, and the captain, who answered to the OIM except when the rig was moving between wells, had already been in Korea for months. They had been monitoring the progress of the rig construction from a temporary field office, the working euphemism for two large shipping containers that had been dropped a hundred yards from the drydock where the Horizon sat.

As the February 2001 launch grew near, more and more managers—*layers* of management, really—arrived to familiarize themselves with the rig's equipment and systems and to set about preparing step-by-step procedures for everything from maintenance of the air-conditioning system to emergency evacuations. The managers of the drilling operations—men with quaint oil field titles like senior toolpusher, toolpusher, driller—were joined by mariners with familiar titles like chief mate, second mate, and not so familiar ones like dynamic positioning officer. Doug's group included technical specialists—the mechanics, electronic technicians, supply supervisors, and subsea engineers who dealt with some of the most important equipment on the rig, the tools that would be lowered beneath the sea to control the well when drilling began. R&B Falcon had given the rig managers a virtual blank check to furnish and appoint their brand-new homes. So many crates of expensive equipment—from new wrenches to reading lamps, flat-screen TVs to office chairs—would eventually accumulate in their cramped work area that movement was nearly impossible.

But they were still a long way from any of that as the bus stopped to drop off Doug and his new colleagues at the crude field office, which they immediately took to sardonically calling "the White House," where they were destined to spend most of their time for the three tedious months remaining until the rig floated. As they acclimated to the strange surroundings, getting used to kimchi at every meal and the way the Korean workers began each morning with mass calisthenics to the beat of American rock songs blaring from loudspeakers, the resident project manager made it clear that until the rig floated, all that any of them could do when they ventured onto the Horizon was watch how the Koreans handled the equipment and take notes.

For much of their twelve-hour shifts in the White House, they were planted in front of computer screens, writing a kind of owner's manual for the new rig and all its equipment. In Doug's case, as chief mechanic, that meant creating procedures for the operation, care, and maintenance of his primary concern—the six huge engines—and all the gritty details: the sewage systems, ventilation blowers, fuel lines, water delivery, and the myriad of other mechanical components required to keep the ship afloat and the rig hands comfortable. If the rig were a Ferrari, Doug and his staff would be the staff of full-time mechanics who not only maintained and repaired the engine, but saw to the air-conditioning, the brakes, the suspension, the power windows, the wiper fluid—everything that moved or sparked from bumper to bumper.

It was all essential, if not always exciting.

For now, though, tapping away at the keyboard in the White House, it was mostly just boring.

At his first opportunity, Doug pushed away from the table he was calling a desk, the computer he'd barely set up, and went out to the rig. It rose 395 feet from the ground to the top of the derrick, the equivalent of a forty-story skyscraper.

He took the temporary stairs that had been plopped down beside the rig by a crane and climbed to the third deck, which was actually the first deck from the ground. He threaded his way through the uniformed workers, an incongruous white giant passing through the crowded construction site, and headed for the engine room. The engines were lined up at the aft end of the third deck, three on the port side and three on the starboard. Between them, and half a deck up, was the engine control room, or ECR, which would be Doug's office. But he went straight for the engines.

There were six diesels, all muscular curves and cast steel sinew, the length of a semitrailer and three times his height. The nearly 10,000 horses they generated served a single purpose—to rotate a generator shaft 720 times per minute, sweeping it through a magnetic field to create 11,000 volts of pure electrical energy. The electricity would surge through the ship's veins, arteries, and capillaries to power everything from the smallest light socket to the big thrusters mounted 125 feet below his feet. The engines were what you'd build if money were no object, gleaming with as-yet untapped power.

The Korean technicians fussed around them, looking like laboratory assistants with clipboards, timers, and measurement devices recording every vibration—every movement—each engine made. As Doug drew near, he could feel the Koreans watching him skeptically, this large, lumbering American with sloping shoulders and thick legs, drawn to the engines as if against his will.

For the first time he felt the urge he would feel a thousand times more, to reach out and run his hand along the diesel's flank, to whisper his admiration. But the next thing he knew, the Koreans coalesced around him in a protective knot and admonished him shrilly. The translation was clear: look but don't touch. Until all the papers were signed and sealed, this rig still belonged to them, and they weren't going to let Doug lay a finger on those engines.

Coming back down the stairs, Doug felt the bitterness of the Korean winter wind. It was so cold. Growing up in California and spending his working years between the waterways of Iraq and the balmy Gulf of Mexico, Doug had never known a high-latitude winter. If he had, maybe he wouldn't have left his cup of coffee in the cold while he sought out his engines—he returned just then to a mug of caffeinated slush. And maybe he'd have packed more than a sweatshirt to stay warm.

Doug had a streak of stubbornness in him, too. He sat there every day typing up his procedures, his body tensed against the cold, shivering and turning a bluish tint. When he had to go outside, he ran from one door to the next, his shoulders to his ears.

A big, strong, moon-faced man had been watching this dance with equal parts amusement and sympathy. He approached with a duffel bag and pulled a pair of thermal coveralls from it with a mangled, four-fingered hand. "Looks like you might need these more than I do."

His name was Jason Anderson. He'd worked oil rigs since his first (and only) summer at community college, when his dad pulled strings to get him a job chipping paint and lugging garbage. It was scut work, the bottom of the ladder, but something about the life aboard a rig grabbed him. Jason got through another year of school, but as fast as he could manage he was back on a rig full-time, moving quickly to the center of the action as a drill floor roughneck. Roughnecks were pushed hard twelve hours a day by the drillers, who weren't always picky about their motivational techniques. With the sun beating down hard and no protection from the wind and rain—coupled with the scuffs and bruises of constant manual labor—most roughnecks had the physical proof of their position after a couple of years. Some remained in that job much longer than it took for the skin of their necks to coarsen, but Jason was smart and motivated. He worked his way up the stations of the rig, the jobs as blunt as their names– shaker hand, pit hand, derrickman. Everyone knew he'd eventually make toolpusher, the drilling foreman. He had a gift for getting people to want to work for him. He was the kind of man who would give up a pair of warm coveralls to a cold stranger in a strange place. It was a gesture that sparked a friendship that would last as long as the Deepwater Horizon.

CHAPTER TWO

OIL AND WATER

1896

Summerland, California

Since a time beyond memory, black streaks of oil and thick veins of tar have seeped from the rocks and sand of the California coast. Native people used it as caulk for their canoes and glue for their tools. Early European settlers dug pits to mine it for building material and fuel. Toward the end of the nineteenth century, oil wildcatters working on the California coast south of Santa Barbara in a place called Summerland realized that the seeping indicated a large reserve of crude oil beneath the surface. As they sunk wells into the ground, they followed the oil field toward the ocean. The closer they got to the shoreline, the more oil they found. When they drilled on the beach itself, the returns were greater yet.

What would you have done?

In 1896, the first offshore oil derrick rose at the end of a pier stretching into the surf. When the well sucked the oil dry, the pier was extended and another derrick built, until finally the water was simply too deep to go any farther.

Eleven years later: a surveyor in east Texas for the Gulf Refining Company, intrigued by the eye-stinging mists that lingered above the dark waters of Lake Caddo, held a struck match over the side of his small, trolling boat and caused a burst of blue methane to flare into the night. Three years later, Gulf Oil floated an armada of barges bearing pile drivers and derricks up the Red River into the lake. Using local cypress trees, they drove pilings into the lake bottom, then built freestanding platforms far from shore and commenced drilling.

Not far away, in the Louisiana swamps, the Texas Company was also having success with pilings and platforms. But building platforms from scratch was expensive, and they weren't easily recycled one well to the next. That meant the company was paying construction crews for lost time when oil wasn't flowing. The financial imperative was obvious. They conceived of an odd plan: they sunk two large barges in the shallow swamp as a base, then welded a platform to support a derrick on top. When the well was depleted, the barges could be refloated and towed off to the next site, where the process would be repeated. The lost time was reduced from more than two weeks to just two days.

Eventually, the entire arrangement was constructed in advance—a drilling platform perched on tall columns attached to two immense pontoons. The platform was towed, floating on the pontoons, to the drill site. Then the pontoons were flooded, and the whole arrangement sank to the bottom, supporting the platform above the surface. When it was time to move to another location, the pontoons were refilled with air and rose to the surface. They were called submersible rigs, and with elongated columns, they could operate in as much as a hundred feet of water.

In 1961, one of these submersible rigs, owned by Shell Oil Company, was being towed in heavy weather. The platform, rocking like the hand of a metronome high above the waves, threatened to capsize the rig. The operator, desperate to lower the center of gravity, allowed water to partially fill the pontoons, which sank below the surface, but not to the bottom. Instantly, the stability of the rig improved dramatically.

A Shell engineer named Bruce Collipp happened to be on board. Collipp had a background in naval engineering—unusual in the oil business, but ideal for the situation at hand. He immediately saw the possibilities suggested by the performance of the partially submerged rig. He came up with a design for a drilling vessel intended to operate at varying levels of submersion, controlled by ballast tanks in the pontoons. The design proved to be so stable when the rig was partially submerged that it prompted an outrageous idea: drilling could commence while the rig was still afloat. This was no longer a clever way to elevate a platform above the waves. It was a kind of ship.

But Shell didn't want to call it that, for fear of placing operations under the thumb of the famously tough maritime unions. So when Collipp was explaining his design to the New Orleans Coast Guard officer, he stressed the idea of a submersible that, nevertheless, operated while only partially submerged. The officer, apparently convinced, merely took up the license application and, under "vessel type," wrote "semi-submersible."

That same year, the National Science Foundation and the National Academy of Sciences launched the first attempt to explore the earth's deep geology beneath the planet's outer crust. It was known as the Mohole Project. Drilling into the earth's inner mantle would have been virtually impossible from the solid surface, where the crust can be twenty-two miles thick. Project designers came up with an audacious strategy: take advantage of some of the deepest waters in the world to get thousands of feet closer to their target.

They hired the experimental offshore rig CUSS I (named with the initials of the consortium of oil companies that owned it) to test the theory that a rig could operate in deep water. Project engineers developed an entirely new system for drilling that used radar and underwater sonar sensors to provide continuous information about the ship's exact position—"dynamic positioning," they called it.

Some garbage scow.

While on board the CUSS I for *Life* magazine, Steinbeck wrote:

> *Success or failure of the daring enterprise hinged upon the ability of the pilot to hold the CUSS at a precise position, as unmoving as if planted in concrete instead of being adrift on a turbulent ocean. To achieve this the pilot used a complex control console designed by Marine Engineer Robert Taggart to regulate four outboard motors, two on each side of the CUSS, which could nudge the vessel in any desired direction. The crucial task was to keep the drilling derrick directly over the hole in the ocean bottom, 2.3 miles below, where the drill turned in the earth's crust. The long string of pipe would bend only so much. If the barge moved more than 1,000 feet from center, the pipe would snap, the drilling operation would be over—and the resulting recoil would jolt the craft and seriously endanger the crew. To avert such a disaster, the pilot watched blips on his radar and sonar screens . . . their pattern told him at any given moment whether and where he was drifting so he could compensate by manipulating motors.*

A 1962 analysis of the experiment concluded: "The system operated successfully for a month at sea and proved that dynamic positioning was not only feasible but entirely practical."

But the CUSS I system, which could still be defeated by unfavorable wind, current, or an inexpert operator, was crude compared to what would follow. In the mid–1980s, when the Global Positioning System was opened to nonmilitary use, semi-submersibles came of age. Minute changes in the rig's positioning were instantly sensed by the GPS unit and fed into a computer, which could make constant and very slight adjustments to the direction and thrust of the propellers, keeping a rig the size of an office park idling within twenty feet of an exact point on the ocean's surface in a gale or raging current. If the seas and winds were calm, the margin of error was no more than the size of a king bed. It was perpetual motion in the service of standing still. Or in the Zen koan of technical nautical language, "under way, not making way."

Geologists had suspected the existence of major subsea oil deposits since before World War II, when they realized that the same geological formations that signaled oil reservoirs beneath the surface of the land would also likely occur beneath the depths of the ocean.

Over the years, estimates of how much crude languished beneath American waters kept rising, up to just under 90 billion barrels today—about 9 percent of the world's estimated total. It was not nearly enough to challenge the dominance of Middle Eastern oil, which makes up two-thirds of the planet's estimated reserve. Nonetheless it was a significant boost to America's dwindling stock—enough to supply all of the country's oil needs for a decade. In the same period, known American petroleum deposits began to pump dry, and Western oil companies' access to foreign oil fields plummeted. Exxon Mobil, British Petroleum, Shell, and Chevron once collectively controlled most of the world's oil reserves, but aren't even among today's top ten. State-owned monopolies, primarily in countries not especially friendly to the interests of the United States—in the Mideast, Russia, China, Nigeria, and Brazil—dominate.

Desperate as Western oil companies were to expand their reserves, all that oil trapped beneath deep water remained tantalizingly distant: harvesting oil from the deep ocean is two to three times as expensive as is producing oil on land. With the price of oil stalled or declining through the 1990s, the companies that specialized in offshore drilling trod cautiously and were reluctant to commission new rigs.

Oil prices bottomed out in 1998, at the 2010 equivalent of sixteen dollars a barrel, the lowest inflation-adjusted price since before World War II. Then they began to climb. By 2000, as most Americans watched in dismay and anger, the price for oil had doubled, and it was on its way to more than quadrupling in ten years. The first rumbling warnings of "peak oil"—a theory that known oil reserves would shrink a little more each year until they had vanished—began to appear in the media. The common assumption was that as oil prices increased, the "pain at the pump," as the headline writers put it, would push the economy toward conservation and alternative, renewable energy sources, just as an oil price spike had begun to do during the short-lived crisis of the '70s.

Back then, high prices had shocked consumers, who cut oil consumption dramatically, which led to oversupply and steep deflation of oil prices. That history lesson made some oil companies wary of moving too fast once the price of oil began to spike again. As it continued to rise, the caution was replaced with confidence and a sense that the world had changed in a fundamental way since 1975, a way that spelled good news for the oil industry. The two most populous countries in the world also happened to have the fastest-growing economies. China and India, with 2.5 billion people between them—more than a third of the world's population—were on the way to tripling their per capita wealth, and inevitably their consumption of energy. Hundreds of millions in China and India, instead of living in huts and riding bicycles, would be moving to air-conditioned homes, buying cars, building factories, shopping centers, flying in commercial jets.

That all takes enormous amounts of energy, and by far most of it would have to come from fossil fuels. Use of renewable energy sources—water and wind power, solar, geothermal, biomass fuels and others—will no doubt increase over the first half of the twenty-first century, but the consumption of oil, gas, and coal will increase even more. At least until we can match the power and convenience of fossil fuels with renewable energy sources.

Addiction to oil and gas is simple: it's very hard to beat hydrocarbons for efficiency. They are on the order of one hundred times more powerful per ounce than the most powerful batteries, solar panels, or wind generators. To generate the thrust necessary to lift a passenger jet, a battery pack would need to weigh tons and be roughly the size of the airport terminal. Not very aerodynamic.

Far from sparking an epic flight from fossil fuels to sustainable energy, the profits from spiraling oil prices ignited a black gold rush, a frantic race to locate and remove the last great undiscovered oil reserves on the planet, which were almost certainly beneath the world's oceans.

And the profits were immense. In the first half decade of this century, the top five Western oil companies pumped their coffers full, netting nearly half a trillion dollars.

Oil company execs who were betting the 2010 equivalent of half a billion dollars on the Deepwater Horizon weren't exactly lying awake at night worrying about the end of oil. Neither were the engineers who devoted whole careers designing the next generations of oil rigs, nor the men like Doug Brown and Jason Anderson who made their living on board, or the millions purchasing shiny new SUVs expecting plentiful gasoline to fill the tank.

No doubt the oil windfall funded its share of diamonds and fine wine, villas on the Riviera and mansions in Connecticut with Mercedes parked regally in the driveways of both. But the unstanched flood of dollars also financed an explosion of technological advancement and the construction of the most sophisticated industrial machines ever built.

CHAPTER THREE

COLD COMFORT

February 2001

Ulsan, Korea

The Deepwater Horizon was to be the first of a new generation of semi-submersibles, the fifth. Previous generations had been inching toward what, in the Horizon, had become almost total computer control of all the rig's critical functions. On a skeleton of steel and concrete, designers had threaded a web of electric cable that branched out across every surface and bristled with sensors that reported back to computers in the rig's control centers. This was the rig's nervous system, and the computers, ordinary PCs, were the brain. The real power was in the software, among the world's most advanced, capable of considering the millions of data points flooding from the thousands of sensors and instantly interpreting and applying that information to make intelligent decisions concerning the rig's performance. In addition to registering changes in position, current, and wind for the dynamic positioning system, the sensors alerted the computer to every broken hatch, clogged pipe, overheated engine, and burned-out bulb, every yee and yaw of the drilling platform, each sniffle and sneeze of the ventilation system. It automatically closed hatches to seal out gas-contaminated air, shut down electrical systems to avoid dangerous sparks or risk of electrocution, and sounded evacuation alarms. The builders boasted that these new systems would make the Deepwater Horizon the safest rig ever floated.

As the February 23, 2001, completion date approached, the assembling crew swelled with the rank and file: the assistants, the roustabouts, the ordinary seamen. Most were experienced hands and knew what to expect from the on-again, off-again life at sea, including the fine distinctions and separations between the subcultures that proliferated on almost all American-staffed rigs bound for the Gulf.

The top managers and the engineers often had advanced degrees to go with six-figure salaries and posh homes, while for the drill-hands like Jason, a diploma from a community college *was* an advanced degree. For the most part, the drill crew came from small southern Gulf Coast towns with names like Pipe Creek, Sandtown, Jonesville, or State Line. They were a hardy, beefy breed who, failing to find substantial work in their small communities, got on with the oil rigs. They lived close enough to the Gulf Coast ports that they could drive the couple of hours to the helicopter shuttles that swept them out to sea. To them, the rugged, monotonous grind of twelve-hour days, seven days a week was well worth salaries of $40,000 to $70,000 or more, which they couldn't possibly make at home. It enabled them to buy cars and houses, Xbox 360s and ATVs. After their exhausting hitches, they rushed back home and made love to their wives, took their kids to baseball games, barbecued with friends and family, drank beer, watched football, hunted and fished. Especially hunted. Jason Anderson had severed two fingers in a childhood bike accident, but when only the middle fingertip could be found, the family asked a surgeon to sew it to his pointer. The boy was damn sure going to need a trigger finger.

Then there was a much smaller group of technical specialists, engineers, and the professional sailors who'd been trained in a little-known network of maritime academies ringed around America's coast. These tended to be northeasterners for the most part, or Left Coasters like Doug, who had to earn the respect of the southerners if they wanted to fit in with the majority culture. If they failed, their lives on board were lived in seclusion; they sat at their own tables for meals and went back to their rooms alone once their shift was up. Here in Korea, though, on the far side of the world, all of them were, at least for a time, Yankees.

The Koreans, the supposed aliens in this story, may have been baffled by much the Americans said and did, but they shared a passion for two crucial items: team building and barbecue. Sometime in December, shipbuilders welded up a turbocharged version of a barbecue grill and presented it with much ceremony, featuring incomprehensible speeches and a long line of bowing men.

The Americans took the gift to heart and installed the grill on the roof of one of their residences, where they threw communal pork-outs every Sunday—and one memorable blowout on New Year's Day, 2001. A particularly frigid wind swept across the roof, which kept them all huddled close around the glowing heat of the grill. The beer was nearly frozen but, as Doug pointed out, still warmer than the air.

The Sunday BBQ ritual—which they would continue, albeit

without the beer and cocktails, once the rig got under way—was a small, good thing; due comfort after the mind- and body-numbing work of typing procedures in the cold hours upon hours through the week. They all felt the release when they boarded the bus back to their temporary apartments in buildings with names they never learned to pronounce. The apartments were pleasant but small, not built for American rig workers, who even by American standards tended to be somewhat oversized. But the lack of elbow room was almost made up for by the abundance of gadgets. In Korea, doors don't have keys, they have electronic keypads; peepholes had long since been replaced by miniature video cameras; and even the toilet seats buzzed and blinked. As some discovered the wet way, the pushbutton nozzle affixed beneath the seat was *not* to clean the toilet bowl. But the gadget many of them looked forward to the most was the radiant floor heat. As soon as they walked through the door, they kicked off their shoes and let the floor warm their soles.

Some crew members found other ways to stay warm.

Brothels in Korea took three different, notably diverse forms: For those who were short on funds, or merely cheap, the prime option was certain barbershops, their special nature marked by having two barber poles out front instead of the usual single one, and a dust-covered barber chair inside. In a dim back room, an attendant told the customer to strip and lie faceup, then placed a hot towel over his eyes and sternly admonished him not to take it off. A story circulated about one worker who ignored the edict and returned to his apartment psychically scarred after catching sight of the paunchy, middle-aged woman with gray pubic hair climbing over him. "I was trapped," he protested to all who would listen, still traumatized. He only perked up later when a naïve young crew member who'd never before been out of the United States, or even Mississippi, happened to ask him if he knew a good barbershop.

"Definitely," he responded, and sent the kid to the place with the two barber poles: "Just do what they tell you."

For those who wanted sex without a blindfold, there was a slightly more posh option: coffee delivery services. A phone dispatcher took your order for what seemed like a very expensive ration of coffee, and a few minutes later an attractive young woman appeared at your door with a cup of joe and a handful of condoms.

Feminine companionship could be found without such a stark quid pro quo. Abundant nightclubs with familiar names, American music, and English-speaking hostesses were packed with pretty girls who possessed good grammar, tight clothes, and Philippine visas stamped "vocalist" or "musician." The surprise was that these girls could really sing or play their instruments. In an attempt to limit the exportation of sex workers, the Philippine government required all women traveling on such visas to demonstrate their musical abilities at the passport office. Those who failed the audition were returned home to grinding poverty, while the talented were put on an airplane to pursue dreams of fame, fortune, or—failing that—an American or European husband.

Instead they wound up singing for rig workers and hustling drinks to men with little interest in their vocal talent. They got a cut of every outrageously priced glass of heavily watered Kahlua and cream they lured a man to buy for them, but an even bigger cut if a customer purchased their time for the entire evening—an evening that began with a little drunken making out in the bar and inevitably progressed to the hotel room bed. The decision how far to go was up to the individual bar girls, but the prospect of finding a rich American husband could be a powerful motivator. The following morning, about a half hour after the 5:30 bus left for the shipyard, there'd be a secondary exodus of young women in rumpled party clothes emerging sheepishly from the crew apartments.

The long days and long nights continued through the winter as the Horizon approached completion. The two mirror-image crews of 126 needed to sail the rig to the Gulf had been assembled, and after months of paperwork, some adventures and misadventures, a few hangovers, and rumors of one or two cross-cultural pregnancies, the documents were signed, the Koreans bowed, and the Deepwater Horizon finally belonged to the Americans. This high-tech cross between a neighborhood, an industrial park, and a factory would become at least as much a home to them as the ones they'd left in California, Texas, Alabama, or New York.

Doug Brown and Jason Anderson were both down-to-earth, practical men not inclined toward demonstrative emotions. But like many of their rig mates, they already felt an affection and pride for the new rig, for its power and sophistication, even before they'd spent a night aboard. But their feelings were tempered by the knowledge that for all the no-expense-spared cost of construction and outfitting, the half decade of design and development, the seemingly bottomless trust placed in their hands—none of it had yielded its owners a single dollar. Before Deepwater Horizon and its crew could begin earning their keep, the rig would have to travel fifteen thousand miles around the southern tip of Africa to a well that was waiting to be drilled a mile deep in the Gulf of Mexico.

CHAPTER FOUR

SEA LEGS

February 2001

Fort Schuyler, The Bronx

When the Deepwater Horizon floated off the drydock into the Sea of Japan on February 23, 2001, Dave Young would have found it ludicrous to think his destiny was linked to a new generation of oil rig.

In just two months he would finally graduate from college with an engineering diploma and a merchant marine officer's license in his hands after six years spent preparing for a career at sea. Yet if Dave could have glimpsed the Horizon creeping toward him from across the globe at roughly the speed that he would soon march across the graduation platform, chances are he wouldn't even have recognized it as a ship.

In that way at least, Dave Young was no different than so many of the men—and an increasing number of women—trained sailors who wound up piloting the self-propelled oil rigs and drillships atop the world's oceans. The path that would take him to the Deepwater Horizon began on the spit of land where the East

River flows into Long Island Sound. The scenic sweep of wide water, anchored by a nineteenth-century stone battlement called Fort Schuyler, is more formally known as the State University of New York Maritime College.

In a sense, Dave, short and tough, supremely self-confident, perfectly represented the scrappy, resourceful, unruly spirit of his college, little known even in its own southeastern Bronx neighborhood. SUNY Maritime offered a military academy–style program, but unlike the naval or Coast Guard academies, there was no requirement that graduates serve in the military. The student body of seven hundred or so seemed to be made up disproportionately of the sons and daughters of local cops or firemen who knew someone who knew someone who graduated from SUNY Maritime with a mariner's license, *and you wouldn't believe what this kid is making.*

Despite being the oldest maritime college in the country and being fully incorporated in the state university system, it remains obscure. Most people hardly know what "maritime" in its title means. Founded after the Civil War to shore up the country's flagging merchant marine ranks, for more than half a century the program had no land-based home—it was conducted entirely aboard a training ship and was in constant danger of closing. In 1934, Franklin Delano Roosevelt, in his last act as New York governor, put the academy securely on the fifty-five acres of waterfront on the eastern edge of the Bronx at the foot of Fort Schuyler, which would lend the college its name.

The average starting salary of Schuyler graduates ($62,500 in 2009) is higher than that of Harvard or Princeton graduates, but students tend to harbor a Rodney Dangerfield–style sense of disrespect. They tell, and half believe, the story about legendary New York builder Robert Moses: He so despised FDR that soon after the president's untimely death Moses moved forward with plans to build a bridge over his legacy just for spite.

It's undeniably true that Moses despised FDR, and that he wanted the Fort Schuyler waterfront to become a public park for, as one ally put it, "300,000 to 400,000 to enjoy every weekend," instead of those "300 to 400 boys." But the Throgs Neck Bridge, which cuts directly over the campus, would not open until 1961, too late to snub FDR but in time to ensure that generations of cadets would be woken in the middle of the night by the backfire of large trucks and that classrooms would be filled with the whirl of traffic helicopters.

Dave fit the bill of the scrappy and resilient underdog type. About five foot eight, with an athlete's balance, he had wary brown eyes that always seemed to be calculating angles. He had the kind of smarts that were rooted in common sense and knowledge of what makes people tick. He was always thinking, always looking to maximize results while minimizing the work it took to achieve them. He constantly carved out time to study, or to work on his boat or his friends' cars, or, most often, to venture off with his crew on a hunt for beer and girls.

Rich Robson was the brains of the group. The son of a New York City cop, he could always manage to ace his exams without ever seeming to study. Matt Michalski, the son of a construction contractor, grew up in a row house in a working-class section of Philadelphia. Matt was the most driven in his shipping classes. He'd grown up envying his father's friend, who had become captain of a containership and managed to retire to a beach house in the Florida Keys at the age of forty. Mark Hanson, also a cop's son, was the wild one, always pushing the group to walk on the edge.

But Dave was the one with the flair for taking the lead. One of his brilliant innovations made him famous on campus in just his first year: He figured out that if he sewed the front collar and tie of his uniform into his SUNY Maritime jacket, he could wake up, pull the ensemble over his head, and walk out of the barracks instantly ready for morning formations.

He got away with it, too. After all, SUNY Maritime wasn't Annapolis. The white shirts and black pants worn by cadets, the traditional uniform of the country's merchant marine, looked similar to the untrained eye, but varied in details as fine as the alignment of belts and the skew of the cap. Uniforms routinely went unironed, shoes unpolished. But still, there were rules, and demerits for breaking them. The college commanders, usually retired military officers, delighted in sneaking up behind cadets who were hatless, or had an untied tie, and writing them up on the triplicate forms they always carried. The pink copy went to the defaulter, so demerits inevitably came to be called "pinks."

Dave had racked up a thick stack of pinks: a nice fistful for not saluting officers or, until his jacket breakthrough, being out of uniform. But most were for being late to morning formation, which was at 0730 five days a week—a tough wake-up call when you'd been alternating nights between calculus books and drinking in Manhattan clubs, not getting to sleep until four or five in the morning as Dave did regularly.

In one period of particularly heavy celebration, Dave and his team of buddies weren't just late for formation. It was spring, the weather was nice, the bars were packed—and so they decided not to show up at all.

Since they all lived on the same floor of the same dorm, and formation was organized by residence, it made a very loud statement when the entire first row simply went missing, a fact that even the numerous upper-class cadet officers Dave had befriended couldn't overlook. This offense was beyond a stack of pinks. It earned them all a summons to appear before the company commander. On the night before they were due on the carpet, Dave and his crew stayed up past curfew, then stealthily sprinted about the campus, flipping the circuit breakers in all four dormitories. In the hours that remained before dawn, they spit-polished their shoes and ironed their uniforms. When the commander came on the nearly empty parade ground, he was met by the front row of formation intact and gleaming in military splendor, standing precisely at attention. A forlorn handful of others who had managed to wake up without the aid of electric alarm clocks blinked in confusion.

It had been a risk for Dave to take such bold action; getting caught in such a stunt would have surely led to his expulsion. But he was always willing to "put his boots on" to run to the aid of a friend. Had he been alone in being summoned to the carpet, Dave probably would have accepted his demerits with his usual cavalier attitude and taken his punishment in stride, but to get his boys out of trouble, this bold move seemed worth the price of failure.

The contrast between the mass of absentees, and the spit and polish of Dave and his crew, worked its magic. The commander was impressed with their obvious contrition and renewed dedication, and condemnation turned to praise. The rest of the class, he intoned, could learn something from these cadets.

Dave had grown up in a split household, spending weekdays with his mom in a leafy, stone-walled section of semirural Connecticut and weekends with his dad on the North Fork of eastern Long Island. The two homes were a three- to four-hour drive away, but only twenty to thirty minutes if you could find a speedboat and roar across Long Island Sound. It wasn't a coincidence that Dave was always oriented toward the water and boating. His dad ran a marina and in his off time raced tiny wooden and fiberglass boats he built himself. Despite their small, 25-horsepower outboards, the hydrofoil racers reached speeds of close to eighty miles per hour and were in constant danger of flipping. The excitement and danger of it was the elder Young's passion—one he shared with his son.

North Fork was a place of contrasts, the wealthy summer homes of Manhattan's elite on one side and the humble ramblers of year-round residents on the other. Dave's dad was in the middle of it all, in every sense. He earned a good living at the marina, but he was far from wealthy. Still, he did have time to spend on his exotic hobby, which was building racing boats to replace the ones that had gotten smashed, or in working on his house. There was always a project going on, and Dave was a gifted and enthusiastic apprentice. Soon he was building the small racing boats on his own. He also built a custom truck painted bright yellow and jacked up on huge tread-heavy tires. He delighted in driving into Manhattan and parking it ostentatiously in front of the bars. He'd jump out of the bright yellow monster truck in one of his shiny polyester shirts and a sunburst smile. Maybe Dave wasn't conventionally handsome, but considering the theatrics, it hardly mattered: In no time the ladies would be orbiting him like planets.

Dave excelled in handiwork and extroversion but was just an average student. His father saw a lot of himself in Dave. And he saw a chance for Dave to achieve a goal that had been denied to him.

Thirty years earlier, Dave's father had attended the Maritime Academy for a year, but had to quit after coming down with mononucleosis, a lingering illness that had made it nearly impossible to keep up with the academy's curriculum. He ended up in the Air National Guard but maintained many of his Maritime connections. And now he realized the school might be a good fit for his son: the rigorous maritime training, the summers spent sailing to ports around the world, the in-state tuition, and the lack of military commitment.

Dave agreed, but he quickly learned that the academy's relatively low entrance requirements didn't translate to an easy ride once inside, especially for those like Dave who enrolled in the school's most rigorous academic program, naval architecture. Dave never envisioned a nine-to-five life behind a drafting table, but he was infatuated with the construction of ships. The physics and mathematics of vessel buoyancy and propulsion satisfied his intense desire to learn not only how to operate ships, but also how ships operate.

An engineering degree can be demanding in itself, but at SUNY Maritime the requirements for earning a United States Coast Guard–issued officer's license more or less doubled the workload. There were also the daily rigors of cadet life to contend with. Underclassmen spent early mornings at cleaning stations in the dorms waxing floors and washing windows, midday in classrooms and labs, and afternoons and weekends behind paintbrushes and chipping hammers on the school's 565-foot, 17,000-ton training ship, the TS *Empire State VI.*

Moored at the dock in the school's backyard, the ship is one of the country's last five-hatch steam freighters—the hatches being steel covers in the deck that open on cavernous storage spaces in the interior of the vessel, from which crates and barrels and pallets of miscellaneous cargo would be offloaded by burly longshoremen, as in the classic Marlon Brando film *On the Waterfront.*

While most steam freighters have long since been rendered into razor blades and rebar, those that remain are crusted with rust and abandoned in ship graveyards around the world. Not the *Empire State.* The grand vessel was built half a century ago but remains in better condition than most ships built in the last half decade. This is thanks to the work of the cadets, who each year beat the steel bulkheads with a million hammer strokes, scale rust with hand chisels, and layer thousands of gallons of paint onto her hull. While Dave was trained by his father in the equipment and technique of modern rust removal—sandblasters, bead blasters, and water guns of such high pressure they can bring the metal down to its purest, whitest steel—at Schuyler, he and his classmates were chained to a tradition of manual labor, of earning their keep. When Dave went without a hammer, he was usually carrying garbage, mucking slop out of bilges, or scrubbing the decks.

It was a demanding program, and not to everyone's tastes and capabilities. Of the enrolling freshman class, less than half would ever make it to graduation. Suitcases and boxes seemed to flow out of the dorms in an unceasing stream. Cadets left who couldn't take the labor, or make the grades, or could not keep the pace of twenty-two-credit semesters; cadets who found the 7–1 male to female enrollment more irksome than they had imagined, or discovered, the first time they set foot on the training ship, that they were desperately susceptible to seasickness.

And then there were those who just couldn't hack the military-style discipline, even the relaxed SUNY Maritime version.

Dave could hack it just fine. For most cadets, working off all those pinks by—what else?—even more chipping of rust and painting was almost unendurable. But Dave would quickly talk his way out of chisel duty into something more skilled and consequential, like repairing damaged fiberglass, a job he performed more expertly than the professional sailors aboard. As his ability made itself known, Dave came more into demand. If, say, the captain wanted wood flooring in his cabin, the ship's carpenter would come find Dave. He'd complete his task hours before his fellow pink-slippers finished chipping their rust, and spend the rest of his sentence joking around with the ship's crew. Soon Dave didn't need a pink slip invitation. He actually found comfort at the end of the day in trading slacks for a boiler suit, dark blue coveralls of the type worn by auto mechanics, and walking down to the ship to assist in repairs and maintenance.

Dave also discovered he had a taste for the school's traditions. In his freshman year he joined the Pershing Rifle fraternity, a military drill society founded in 1891 at the University of Nebraska by an obscure second lieutenant named John J. Pershing, later to become General "Black Jack" Pershing, who led American forces in World War I. At SUNY Maritime it was a work-hard, play-hard club with its own hideaway in the neglected basement of a campus building. Dave liked the exacting standards to which the fraternity members held themselves, their pressed uniforms, the synchrony of their parade march, and their sense of self-discipline. He also liked doing Jell-O shots in the rank basement with his frat buddies. It was the latter—strictly against the rules—that eventually prompted the administration to shut the frat down in Dave's senior year.

But despite his unruly streak, and the fact that his bright yellow monster truck took up three spots in the dorm parking lot, the campus brass embraced Dave (to the point of letting him stash the truck on the football field access road, conveniently opposite his dorm) and Dave embraced the school—especially its tradition of sending all students for summer-long training cruises on the *Empire State VI*.

In May, the whole college would steam across the Atlantic. The cadets did everything: loaded the supplies, mopped the decks, stood the watches, steered the ship. Every day at sunrise and sunset they were required to use their school-issued sextants, instruments from the epoch of sailing ships in the era of Global Positioning System satellites, to read the ship's exact position from the alignment of the stars. For a grade.

Dave's crew figured there was a better way, a way that didn't involve waking up before dawn. They knew that at any point in the day, if they could sneak up to the bridge, they could find where the ship had been at sunrise from the history in the GPS. But that wouldn't do them any good—it was like having the answers to a long-form math exam, one in which you were required to show your work. They needed to show the astronomical alignment that, in theory, had led them to the ship's position.

In the quest for more sleep, they pulled consecutive all-nighters, trigonometry books spread out around their cabin, poring over the formulas until they got it: a way to reverse-engineer the results from the pilfered sunrise position. From then on, they slept through the morning observation. When they got up, feeling refreshed and pleased with themselves, they found a reason to be on the bridge, created a diversion, got the sunrise navigational fix. Then they'd work their mathematical magic and voilà, the exact astronomical alignment above the ship at the moment the sun peaked above the horizon. They charged their shipmates five dollars apiece to replicate their work.

When they made port, in Bermuda, Barbados, Athens, Tenerife, Lisbon, Malaga, Hamburg, Portsmouth, Naples, they got a different kind of education. The school catalog called it "exposure to international cultures." They quickly learned that it could soon become overexposure. The locals were happy to see them when they arrived—young men would offer to show them the town, young women wanted to practice their English on them—but by the end of the week, as often as not, they'd board the ship running.

There's a maritime saying that in rough weather, a sailor needs one hand for himself, and one for the ship. So when any Schuyler cadet got in over his head in a crowded Lisbon bar, talking to the wrong woman within earshot of the wrong man, he could yell "one hand!" The streets, perpetually filled with shipmates, would empty at the call, and a dozen guys from the Bronx would come surging through the door.

The SUNY Maritime cadets discovered the more literal meaning of "one hand" on the summer training cruise of 1996. One beautiful morning the sunrise was particularly colorful—the entire western horizon glowed red. Despite the old nautical saw "red skies in morning, sailor take warning," few of the officers, some of whom were veterans of winter North Atlantic runs, Alaskan swells, and transits through subarctic oceans, were worried.

But the ship was heading toward a monstrous storm with hundreds of cadets aboard. One of the few who saw what was coming was the senior deck training officer, an eccentric nautical sciences professor named Gregory P. Smith, forever referred to by his initials, GPS. His ability to grab a sextant, shoot a sun line, and calculate it without looking at tables seem in retrospect bizarrely predestined. Now Smith looked at the sky and the barometer and realized what was bearing down on them just as the digital weather forecasts began to spit out of the ship's fax at a fever rate.

To the cadets' amusement, Smith began tying down all his belongings as if the ship were about to be turned upside down. He suggested they do the same. He urgently shared what he claimed to be the only true cure for seasickness: a strict diet of diluted lemon juice and saltine crackers. The cadets laughed among themselves at what they considered a craggy professor's melodramatic concerns.

But Dave Young wasn't put off by eccentricities. In GPS he saw someone with smarts and hard-won experience, and he knew that when someone like that issued a warning, he needed to heed it.

While most on board continued their daily rituals, Dave took proactive measures. He packed his belongings in his foot locker, then tied it down with heavy manila rope and stashed a supply of lemons and saltines. His friends, who had rarely seen Dave worried about anything, soberly took note and did the same.

Then, like a hammer, the first wave hit, a swell the size of an office building. The wind worked itself octave by octave into an unceasing wail. Soon bodies, desks, magazines, footlockers, pillows, heavy tools—everything aboard that had not been securely tied down to welded steel—shot through the air. The walls caught the mess as they turned into floors and ceilings, the ship rolling on its port side, pausing as if to catch its breath, then rolling on an equally merciless angle starboard. One particularly naïve or reckless cadet had ignored the warnings to such a degree that he was casually running on a treadmill when the first wave hit. On the upward roll, his jog became a desperate scramble up Everest's peak, which, on the downward roll, transformed into a toboggan ride to hell. Sick cadets vomited on decks, creating a dangerous game of summer slip-and-slide, and prompting one extra-large cadet to skid down a hallway and plunge through a wall.

For days, work ground to a halt. Doors leading topside were dogged shut to keep water out and cadets mad with seasickness in, for fear they'd stumble onto the deck and be swept away by a towering wave. Dave alternated between nights spent in bed with his life jacket and Gumby suit (a wet suit designed to totally encapsulate you in neoprene), wedged beneath his mattress to pin him comfortably to the wall, and days spent aboard a lawn chair he'd suspended from the ceiling. As the ship pitched wildly, he swung like a pendulum high above his shipmates, smiling widely, looking serene and slightly unworldly up there, Buddha-like.

But after a few days of enjoying the relative safety created by their preparations while their shipmates retched and rolled miserably in their berths, the boys got bored and decided to leave their secure zone and venture out. First stop was the bow store, a gymnasium-sized space that, due to its distance from the ship's center of gravity, gyrated most dramatically in the waves.

Dave waited until the bow reached the top of a wave, then jumped just as the floor fell beneath his feet and the ceiling plunged toward his outstretched hands. Grabbing a lattice of pipes on the ceiling, he held tight until the bow began to fall again, cushioning the twenty-foot drop back to the deck.

He was just getting started. Dave, Matt Michalski, and Rich Robson decided that for the next trick they would climb monkey island, the topmost deck, so named for the abundance of antenna and halyards that any young monkey would dream of playing on. Going up by the outside staircase, exposed to the waves crashing on deck, would be suicide, so they walked through the dark heat of the ship's engine room, then climbed a ladder inside the ship's smokestack—just feet from the evaporating steam and burnt smoke of the ship's boilers. Luckily, the hatch at the top was one of the few that had been left unsecured—no one else had wanted to make the climb to secure it. They threw it open and pulled themselves onto the flat deck. Huddling behind the steel protection of the mast as the storm thrashed the ship, Dave and Matt and Rich stood against the weather feeling fearless in each other's company.

Five years later, after he had finally subdued the engineering curriculum, Dave Young sat beneath a cloudy sky, surrounded by the protective stone walls of Fort Schuyler and bored out of his mind. His butt went numb as the graduation ceremony droned on. He and his fellow cadets were dressed in their "salt and peppers"—white nautical dress shirt and black pants—listening to the speeches, just waiting for the moment when it would be over. He was two years late to this ceremony, but he'd beaten the odds. Three out of four who began freshman year with him hadn't made it at all, and most of those who had made it had opted for easier tracks of study. His degree had been a tough get. Still, he knew his indifferent academic record was not likely to entice the leading naval architectural firms to knock on his door. Nor did Dave plan to make any effort to solicit their invitations for employment.

During his summers sailing the North Atlantic, Dave had been bitten by a bug that had ruined the higher aspirations of countless mariners before him: He had fallen in love with the independence, camaraderie, and excitement of a life at sea.

As Dave's mother, brother, and father all sat dutifully in the family section to the side of the stage, cadet after cadet proceeded up the stairs and across the platform from right to left, pivoting for the sheepskin handoff and breaking for daylight. Dave could only curse his surname, which had doomed him to be one of the very last. But finally his turn came. He grabbed the rolled diploma and stepped down onto the beautiful lawn. He faced east, into the salty spring breeze pouring over the pentagonal stone walls of Fort Schuyler from Long Island Sound, and looked into the future.

Dave was thinking ships, sailing the seven seas, that kind of thing. In his six years at Maritime, not once had anyone ever suggested a career on offshore oil rigs.

CHAPTER FIVE

KING NEPTUNE

May 2001

The Indian Ocean

As Dave Young grasped his diploma, the Deepwater Horizon was halfway around the globe, sixty days into a seemingly interminable maiden voyage.

It hadn't helped that the Horizon's derrick, rising 320 feet above the water, meant that it was too tall to fit under the Suez Canal's Mubarak Peace Bridge. The crew faced a course that would wind fifteen thousand miles around the southern tip of Africa and bring them to the Gulf of Mexico.

Slowing the journey further was the rig itself. Many earlier rigs had been able to take advantage of a remarkable marine technology that came of age in the very shipyards that built the Deepwater Horizon and its ilk. Called heavy-lift transports, they were essentially gigantic seagoing flatbed tow trucks. Using the same ballast control principles that worked in the Horizon's pontoons, the transports were able to submerge their long, flat, low-lying cargo deck thirty feet below the surface, allowing a rig to float into place above it. Then the heavy-lift ship blew out the ballast and rose up in a gushing fountain of displaced water, lifting the entire rig from the sea.

It was an expensive operation, but the advantage was huge: It could carry a rig at three times the speed the Horizon could make on its own.

And now more than ever, speed was essential. The Horizon was no sooner out of the drydock than it was signed to a three-year lease to drill in the Gulf of Mexico for British Petroleum, the fourth-largest corporation in the world. BP, once partly owned by the British government, was now a publicly traded multinational corporation that operated 22,000 service stations in more than 80 countries and produced nearly 4 million barrels of oil a day. In 2000, BP logged annual revenue of $148 billion, and that number would double over the next decade. The company had successfully developed offshore oil fields in the North Sea and was now a key player in the race for oil in the Gulf of Mexico. To lease the Horizon, it was willing to pay the 2010 equivalent of $350,000 a day for the bare rig, an amount that would only grow over the years—to *half a million* dollars a day.

The agreement came with one overwhelmingly significant proviso: If the Horizon wasn't drilling, BP wasn't paying. Every day the Horizon crawled across the ocean toward the Gulf, Transocean was out a half million dollars. So cutting the crossing by two-thirds would translate to a savings of tens of millions of dollars.

But the Horizon had to make the crossing the old-fashioned way. Ironically, it was the rig's ability to propel itself that was the problem: The eight thrusters hanging beneath its pontoons made hitching a ride on a heavy-lift transport impractical. Disassembling the thrusters that had just been assembled would be troublesome and time-consuming. Besides, Transocean engineers thought they had an alternative: The Horizon would hook up with a tow ship that would pull as the Horizon pushed, theoretically doubling the rig's maximum solo speed of only 4.6 knots.

Theoretically.

The Horizon motored on its own power through the China Sea to Singapore, then out into the Indian Ocean, where it rendezvoused with the tow ship for the longest leg of its journey, the six-thousand-mile trek to Cape Town, South Africa. The mariners on the Horizon worked with the tow ship's crew for a week. They didn't really know what to expect, as they had never worked with the particular drag coefficients of the pontoons and the thrusters, which projected nineteen feet below. They ran through all possible combinations of thruster direction and throttle, but all they accomplished was to burn more gas. They adjusted the cable connecting the two vessels, trying to find the perfect length, one that would minimize the yo-yoing between slack and taut, and ensure that both ships crested the waves at the same moment. But nothing helped. Every inch, every ounce of the Deepwater Horizon had been built for stability, not speed. That was its deepest nature, and it wasn't budging. All the tugging and engine revving in the world couldn't change that, a strength of its industrial character that its crew noted with something akin to pride—even as it condemned them to months at sea. There was nothing to do now but ride it out, cut the tow ship loose, and go the whole route at the Horizon's natural ambling pace—which meant that the voyage they'd hope to make in under fifty days would take ninety-eight, costing the company millions of unbudgeted dollars.

For the crew, it meant a suspension of life as usual, the only long stretch in its history where the Horizon would be more ship than rig. Used to having helicopters fly them to the nearest airport and connections home every few weeks, the crew would have to get used to the idea that until they arrived in Cape Town, they'd have no more options than the crews of nineteenth-century sailing vessels—and even less available speed.

Of course, they wouldn't be drilling, either. But that didn't mean they weren't busy. In fact, the sailors on board were more occupied and urgently necessary than at any time in the rig's life, completing half a circumnavigation of the world, dealing with fine points of navigation, winds, currents, propulsion, and all the unexpected contingencies they'd trained for years to meet. Doug had a blessed stretch in which to bond with his engines and work out all the remaining kinks in the rig's other machinery, and in his own crew of mechanics. And the toolpushers, the drilling foremen, were like nineteenth-century gunnery captains who, with no enemy within a thousand miles, nonetheless exercised the gun crews night after night, urging them on to faster and more accurate practice, so that when a foreign sail did appear on the horizon, they'd be ready, and deadly.

In the Horizon's case, of course, they weren't running cannons out gunports, but conducting performance tests on the cranes and drilling equipment, tuning up the blowout preventer and chasing all the bugs out of the rig's software. And their "enemy" wouldn't bear up on a stiff wind, but would be waiting for them at destination's end—if there was an end—in the form of a BP test well where the Horizon and its crew would have to prove themselves before they could go on to their first working assignment and begin to earn their keep.

As the crew busily attended to their varied pursuits, the divisions between them that had been largely ignored in Korea began to peek through in small ways. The drilling crew, anticipating its deepwater destiny, might pester the mariners, constantly asking, "What's the water depth here?" The mariners, concerned with winds, currents, and making their way across the *surface* of the ocean, couldn't care less once they were far enough from land not to worry about grounding the rig, and didn't have equipment to calculate such great depths in any case. They'd react to the inquiries as if they were dealing with annoying kids in the backseat constantly piping up with "are we there yet?" Even something as prosaic as the derrick lights might become proxy for a low-grade culture war. The drilling crew would turn them on, and the glare off the sea spray would blind the watch-standers, those constantly on the lookout for obstructions floating in the dark water. When the crew complained that lack of lights was creating a safety hazard on deck, the mariners told them to carry a flashlight, that hitting another ship because they couldn't see was a far more significant safety concern. Inevitably, it would take one of the mariners who had been around, who knew how to translate in a way Yankees and southerners and even Left Coasters would all understand. The conversation went something like this:

> INTERPRETER: Pretend this is your pickup truck, and your kid, sitting in the backseat, turns on the interior light; makes it kinda hard to see, right?
> DRILLER: Yeah, but these are *outside* lights!
> INTERPRETER (PATIENTLY): Okay, think about fog. Would you turn on the high beams of your four-by-four in fog? Now think of the sea mist as small particles of fog and the derrick lights as giant high beams . . . see the problem?

That did it. The drill crew not only complied but in the weeks ahead would eagerly search the decks after each sunset looking for rogue lights on the derrick to switch off, a recognition that they were, in the final analysis, all in the same boat.

Six hundred miles off the southern tip of India, the Horizon crossed the equator. Jason Anderson had been here just the year before as one of the crew sea-testing an unfinished rig, Transocean's Cajun Express, which was being towed from Singapore to Grand Isle, Louisiana, where it would be completed. Shortly after the journey, Jason left Transocean for a job with R&B Falcon and ended up here, about to cross the equator for a second time. So he knew what was coming—an ancient and abundantly bizarre initiation called the shellback ceremony.

Shellbacks like Jason, those who had been initiated on a previous voyage, took control of the deck as the crossing neared. The highest-ranking shellback became King Neptune, in this case a wild-man electronics technician named Gene Frevele, whose coils of brown hair streamed from under a tinfoil crown embossed with some kind of crustacean. He ruled over the proceedings with a trident of welded metal rods. Jason, an irrepressible extrovert, enthusiastically played the second lead in this production, as the "Sea Baby." He wore the headband of a welder's face mask trimmed with colored rags that hung down like hair. His shorts had been ripped into rags and his work shirt cut off below his chest, leaving exposed in all its glory his hairy melon of a belly—which he had smeared with grease. He loved his opportunity to ham it up for the new kids, the "pollywogs," who, as the tradition demanded, would be subjected to all kinds of imaginative humiliation and abuse before they could emerge from their embryonic state and join the society of shellbacks.

In the navies of previous centuries, the initiation ceremonies could be brutal. Pollywogs were covered with filth, forced to eat noxious substances, beaten with boards and salt-stiffened ropes, sometimes even tossed overboard and dragged in the surf. Injuries were common and deaths not unheard of. As late as 1995, in a shellback ceremony captured on video aboard an Australian submarine, a pollywog was sexually assaulted with a stick.

Naval regulations have been instituted to curb abuses, but on commercial vessels, individual traditions still determine the nature of shellback initiations.

The guiding ethos on the Deepwater Horizon could be summarized as "good-natured gross-out." The conspiring shellbacks had been saving the food waste from the compost barrel in the mess hall for two days, which they ladled out of noxious-smelling pots and poured into an improvised wading pool. The blindfolded pollywogs had to "drink" the brew (they spit it out) and crawl through the slops, then rinse off in a tub of yellowed oily water the shellbacks called "whale piss."

For their final act of obeisance, the 'wogs were led, still blindfolded, to Baby King Neptune. They were ordered to kneel, whereupon Jason magnanimously accepted the initiates into his kingdom by rubbing their faces in his Vaseline- and food slop-encrusted belly.

When it was over, the slime washed off in the shower, but the bond—of the ceremony and of the shared, interminable crawl across a vast and empty ocean—remained.

The jovial mood wouldn't last long. Just three days later, one of Doug's motormen, a good-natured man named Jack Parento who had robustly played a pirate in the shellback festivities, with a red bandanna and black eye patch, was on shift when he began complaining about a bad case of heartburn. He asked if he could go see a medic, and of course Doug let him go. An hour or so later, the shift electrician found Doug and said, "Did you hear about Jack?"

"What do you mean?"

"I just passed by the medics' office. They had Jack on a bed with wires all over his chest. They said he'd had a heart attack. The medics gave him CPR and brought him back. He's stable, but in bad shape."

They were still too far from anywhere for a helicopter to evacuate Jack. In what they all knew was probably a futile gesture, the captain ordered all thrusters to go full out.

That night, Jack suffered a second heart attack, and could not be revived.

It hit the crew hard. Conversation all but ceased as everyone went through the motions of rig life that night. Doug's shop was especially devastated. Jack had been the joker, always the guy to get everyone loose when things got tense. The coffee cup he'd used every day—featuring a girl in a bikini with cleavage that commanded attention, beneath which was written, "Watch my back"—just sat on a desk in the engine control room, reminding them all how one moment you could be alive, laughing and cutting up in a pirate outfit, and the next, gone.

It made them all feel even worse that they were still weeks away from any place where Jack's body could receive proper attention. There was no choice but to clear out the food from one of the cold storage lockers and store the corpse there until they reached port—still thousands of miles distant. The weight of sadness and the unsettled presence of their rig mate's inanimate body always in the back of their minds, the incremental creep of progress, not to mention the imminent depletion of their tobacco reserves, began to threaten sanity. Every day, for weeks it seemed, the question "how many days left" would bring precisely the same answer.

But even the longest journey ends. For Jack Parento, it ended on the island of Mauritius, off the eastern coast of Africa. The crew was not permitted to depart the rig (save for one crewman with an impacted tooth), but the company had flown Parento's widow in to the island, then landed her on the rig by helicopter. The entire crew participated in the somber memorial service on the rig deck, after which the body was shipped home, while the Deepwater Horizon rounded the Cape and continued to crawl across the Atlantic.

It was late spring when they arrived in the Gulf. The rig latched up to the BP test well and the crew took it through its paces. Some rigs had required months of constant adjustment to pass. The Horizon, having worked all the kinks and bugs out of its drilling systems in the long passage, proved itself in a matter of weeks. Nearly three years after its construction had been commissioned, the Horizon was finally ready to do what it was made to do—but it would not be doing it for R&B Falcon.

The Deepwater Horizon had been a bold addition to what was already the largest fleet of deepwater offshore rigs in the world. Those physical assets, and the great debt that R&B Falcon had incurred in manufacturing them, had driven the company deep into debt, making it a tempting acquisition target for the slightly larger, fiscally sound offshore drilling company Transocean Sedco Forex. As its lengthy name suggests, the company was itself the product of promiscuous mergers that began in 1953 when it acquired the Offshore Company out of Birmingham, Alabama.

Transocean, as the new merged entity would come to be called, had announced its intentions to buy R&B Falcon in a complex $8.8 billion deal months earlier. But the formal document of transfer was not signed until mid-August.

When the crew got the official announcement, Jason, whose first jobs had been on Transocean rigs, told a buddy, "I left them sum-bitches to come to R&B Falcon and now here I am working for them again!"

The truth was, though the two companies had very different policies and styles—from the way hitches were timed to the role mariners played in day-to-day operations, and a hundred other things—the Horizon would be considered a legacy rig and continue to operate with the strong influence of its original R&B Falcon policies. Change would come so slowly that even years later, offshore veterans would be able to walk onto the Horizon and know which company had built it.

But Transocean's culture was powerful, and it would soon stamp itself indelibly on the Horizon. Formally based in Switzerland (for tax purposes), Transocean ran its Gulf of Mexico operations out of offices in Houston, and it cultivated a can-do cowboy swagger as a company that could accomplish the near impossible in the new frontier of ultra-deepwater drilling.

"We're never out of our depth," was the corporate motto.

CHAPTER SIX

MACONDO

Miocene Epoch, 20 Million B.C.

The Mississippi Canyon

Life, no matter how abundant, ends in death. Every death, no matter how small, is significant. That's the larger lesson in the deposits of oil and natural gas that lie buried within the earth. Deposits like the one we now know as the Macondo Prospect.

It formed during the Miocene geologic epoch, somewhere between twenty million and ten million years ago, in the depths of an ancient sea, fed by giant rivers—now called the Mississippi and the Red—that drained the nascent North American continent as it drifted away from Europe and Africa. With the Atlantic Ocean slowly growing larger, the nearly enclosed sea that would become the Gulf of Mexico remained remarkably unchanged. The giant rivers that fed nutrients from the eroding land into its warm confines created an explosion of life, most of it microscopic. Contrary to popular myth (and oil company logos), dinosaurs only ruled certain terrestrial neighborhoods of earth—and their corpses didn't turn into oil. In many ways the most significant impact on the planet has always been from the smallest life-forms, not the largest. Microorganisms produced the oxygen that transformed the earth's atmosphere. And microorganisms are largely responsible for the formation of oil and natural gas deposits, the burning of which is altering the atmosphere once again.

At any given moment, microscopic ocean life accounts for a larger number of living organisms than there are stars in the universe. It's been that way for three billion years. Every drop of water in the ocean contains more than a million microscopic organisms. While alive, most of these organisms employ ingenious strategies to keep themselves near the ocean's surface, where they can transform sunlight into food via photosynthesis. They ride the currents, take advantage of the wind and wave-induced turbulence, propel themselves with tiny whiplike paddles, or even create their own buoyancy by pumping lipids between their cellular membranes. But when they die, the quest for sunlight ends and they drift into the darkness of the deep, to the bottom, where they collect and decay by the trillions of trillions.

The balance between collection and decay is crucial. Certain conditions, the ingredients the Gulf has always offered in abundance, kick the bloom of microscopic ocean life into high gear: warm water, upwelling currents, and the inflow of nutrients. In combination, they spark an orgy of creation and a spike in the microbial population. All that life brings on an avalanche of death. The dead cells rain down in a blizzard—it's actually called "marine snow"—and accumulate on the bottom faster than they can decay. Compressed and covered by the sediments pouring in from the rivers, the thick mat of sludge is cut off from oxygen, which over time might have burned the biomass away. Instead, the accumulated weight keeps pushing the organic matter deeper into the earth. The heat and pressure of the earth's core begin to crush and cook the molecules themselves, squeezing them into a series of ever simpler forms until, after many millennia, they reach their ultimate simplicity. Crude oil. Natural gas. Or methane, whose molecular composition is elegant simplicity itself, a single carbon atom surrounded by a pyramid of four hydrogen atoms.

Now much lighter than everything surrounding them, the liquid hydrocarbon reservoirs exert a powerful upward pressure as they struggle to rise to the surface, a geological version of CO_2 bubbles in a glass of Coca-Cola. They migrate through tiny cracks and fissures or slide up along the inclines separating layers of rock until they reach the surface. Scientists estimate that every year 500,000 to 1.5 million barrels' worth of oil and natural gas seep into the Gulf of Mexico—that's about double the range estimated for the spillage from *Exxon Valdez.*

In geological time, most oil eventually evaporates. Some geologists believe the Ohio River valley may have once had oil deposits as extensive as those in the Middle East. Now, of course, they are long gone.

But as the hydrocarbons slide along the underside of impermeable barriers toward the surface, sometimes they get stuck. They slide up a slope made of nonporous rock or compacted salt deposited by shallow ancient seas, seeking higher ground, only to find the incline has become a trap. These traps are dome-shaped deformations, inverted cups where hydrocarbons slowly collect, eventually forming vast reservoirs of oil and gas. The Gulf of Mexico happens to have a lot of these upside-down cups of oil.

In the past two decades, ships towing three-and-a-half-mile-long cables studded with sensitive underwater microphones (called hydrophones) have made carefully charted sweeps of the North American continental shelf. Periodic bursts into the water with an air gun or a charge of dynamite create seismic waves that penetrate the sea bottom. As the waves hit each new obstacle—a bank of mud, an outcropping of limestone—a portion of the wave energy is reflected back toward the hydrophones. Mud reflects differently than limestone, and limestone reflects differently than salt deposits or impermeable rock. The techniques are not new. Seismic calculations were first made by engineers during World War I to triangulate the positions of large enemy guns. But the equipment developed thereafter was so sensitive, and the computers analyzing their data so powerful, that they allowed geologists to create three-dimensional maps of structures thousands of feet beneath the earth's surface.

In 2003, after months of seismographic trolling, the sound shadow of one of these domes appeared on a study of an area forty-one miles off the southeastern coast of Louisiana in the middle of the Mississippi Canyon, a five-mile-wide undersea ravine that runs along the Gulf bottom for seventy-five miles. The resulting charts of the survey looked like a series of Rorschach tests, and what British Petroleum saw in them was profit.

Five years later, in March 2008, BP bought the rights to drill in what was officially designated Block 252 of the Mississippi Canyon in the United States' exclusive economic zone of the Gulf of Mexico—or actually leased them, since all oil rights on the outer continental shelf, which extends two hundred to three hundred miles from the coast, belong in perpetuity to the federal government. For the rights to explore for oil on the 5,760-acre block of ocean bottom under five thousand feet of water, they narrowly outbid five competing companies by offering $34 million.

In order to maintain secrecy about their new lease, and perhaps with a lingering romantic sensibility from the wildcat days, oil companies designate their prospective sites with code names. BP saw an opportunity for taking care of corporate business in the naming rights. Who hasn't dreamed of bestowing an everlasting name on a continent, a mountain, a star? Why not an oil well? Naming rights for Block 252 were made the prize in a company United Way fund-raising contest. The winner, a BP employee with a literary bent, came up with the name Macondo, after the fictional town created by Gabriel García Márquez and the setting of his masterpiece of magical realism, *One Hundred Years of Solitude.*

In the novel, Macondo starts out as a speck of a town in the middle of the jungle, then expands physically and culturally until it is a dynamic but deeply flawed city whose citizens fall prey to their own greed and begin to take moral shortcuts. Macondo's promising beginning succumbs to a series of plagues and wars, until finally it is blown off the face of the planet by an explosive windstorm. In a final irony, the citizens of Macondo have been warned, in writing, of the tragedy to come, but the warning has been written in a language nobody is able interpret until the final moment, when it is already too late.

For those looking for ill omens in retrospect, the name couldn't have been more tragically apt. Nor could the initial choice of rig to drill the well.

BP chose the Transocean-owned Marianas, a twenty-four-year-old semi-submersible with a long history. The Marianas had been destined to brush up against disaster from her 1979 birth as the MSV Tharos at Mitsubishi Heavy Industries' shipyard in a city synonymous with catastrophe, Hiroshima, Japan. She was designed to fill what the offshore industry perceived as a worrisome gap in its plan to establish oil-producing platforms ever farther from shore. The Tharos and others like her were built not just for saving lives, but also to save the environment by performing well-kill operations that could forcefully bring an uncontrolled well blowout to an end. She was outfitted with what was then state-of-the-art technology: a dynamic positioning system to maneuver her into position alongside a burning rig, an enormous gangway that could activate off her side to give survivors a dry means of escape, and monstrous fire cannons capable of shooting 40,000 gallons per minute a distance of 240 feet, powerful enough to blow a man off the deck of a nearby rig or worse. Get too close, and they could cut a man in half.

The Tharos also carried multiple fast rescue boats and its own Sikorsky S–76 helicopter capable of plucking twelve men at a time off a burning rig and transporting them to the Tharos's ninety-bed hospital outfitted with all the gear needed to sustain life, including an operating room and patient monitoring facilities. She cost a then unheard-of amount, exceeding $100 million. The Duke of Edinburgh called her "the most expensive fire engine in the world."

Not for long. In July 1988, the Piper Alpha platform was at work in the North Sea doing what she'd been built to do—tapping into existing wells, separating the oil from gas, and pumping both cargos through undersea pipelines to tanks 128 miles away on the Scottish shore. The disaster was set in motion by a series of unlucky coincidences and a mundane human error. The rig was undergoing an upgrade, which its managers decided to complete while pumping continued, rather than absorb the high cost of shutting down. A small part of the work involved replacing an old valve on a backup gas pump with a new one. The technician's shift ended before the work was finished, and his replacement was busy with something else. The maintenance worker filled out the paperwork that should have warned everyone on the rig not to operate the pump under any circumstances, but it got lost. That evening, the primary pump went down. When the backup was started, gas poured from the valve vent. A random spark ignited an explosion, which triggered a cascade of increasingly violent secondary explosions, fed by thousands of gallons of crude oil and natural gas.

The blasts took out the main control room, the generator, and the power distribution system, and also destroyed the one chance of fighting the fire—the deluge system, a curtain of lifesaving water. The raging fuel-fed fires made launching lifeboats impossible. Some risked the hundred-foot plunge into the fire and ice hell of the frigid North Sea, now covered with flaming oil slicks. But almost half the crew mustered in the large accommodation area to await evacuation by the Tharos's helicopter then located only a thousand feet way. The helicopter never came. The fire and smoke made landings impossible. Most of the crew burned to death or were overcome by smoke and toxic fumes. Nor did the Tharos's water cannons do much good. The continual replenishment of explosive fuel made fighting the fire a losing battle. And the Tharos's huge gangway failed at the critical moment, just as it started to extend its arm of safety. The rig's crew was close enough to choke on the smoke and feel the heat of the huge fireball, to see the burned faces and charred bodies. Some worked for thirty-two hours without a break, only to watch helplessly as the platform disintegrated.

So confused was the Tharos's crew that the first man it saved swam unassisted to her pontoons and climbed an external ladder to safety. It was only after he told Tharos hospital staff who he was and where he came from that they realized an actual survivor was on board. Of the 226 men aboard the Piper Alpha that day, only fifty-nine survived, many with severe burns. Few of those had been rescued due to direct assistance of the Tharos.

While vessels like the Tharos often did save lives offshore, the Tharos would be remembered mostly for her failure. The industry could have taken the lesson as an opportunity to improve MSV design. Simple improvements, like adding a winch on the helicopter capable of dropping a basket down to a burning rig's deck without the helicopter having to land itself, could have had a sizable impact in the response to future emergencies. Instead, the rig's failure killed the idea of prepositioning emergency support vessels near offshore oil fields, making all future offshore oil workers reliant on distant helicopters and hospitals. And the residents of nearby shores would have to resign themselves to waiting precious days for the arrival of equipment capable of fighting blowouts and stopping the uncontrolled flow of oil into surrounding waters.

The Tharos had changed owners as well as names and seemed destined to take a final voyage to the shipbreakers when, in 1996, she was bought by Transocean Sedco Forex for conversion to a semi-submersible drilling rig capable of being moored in seven thousand feet of water. The company removed the hospital, firefighting equipment, helicopter hangar bay, and even the thrusters to prepare her for her new life. All she needed was a new name to go with it, one appropriate to her role in the company's bold move into ever deeper waters. Transocean sponsored a "name the rig" contest among its employees, and twenty-one-year-old Nora Dossett claimed the honors by coming up with Marianas, in honor of the world's deepest ocean trench.

If BP executives were even aware of the macabre portents of their choice of code name or drilling rig for their new lease in Mississippi Canyon, there was no particular reason for them to fear they were tempting fate. The Piper Alpha disaster was two decades in the past, while the Tharos had been remade and renamed. And the town of Macondo was just a fiction, after all. Plus, by the daunting standards of deepwater oil drilling in the Gulf of Mexico, the Macondo Prospect was not especially challenging. The plan called for the prospect to be drilled in 4,992 feet of water, to a depth 14,569 feet below the ocean floor. BP geologists had identified two likely hydrocarbon reservoirs in sandstone formations, the most promising at 13,319 feet and one a thousand feet farther down, close to the bottom of the proposed well. At 5,000 feet, the ocean depth would be only half that of the record 10,000 feet in which Chevron and the Transocean rig Discoverer Deep Seas teamed up to successfully drill a well in 2003. Likewise, the 14,569-foot penetration beneath the ocean floor was less than half of the 35,000-foot record drilled by BP just a month before the first drill bit hit Macondo. For the record-breaking effort, BP had also used a Transocean rig: the Deepwater Horizon.

Along with its relatively modest depth, Macondo Prospect would be a straight shot from the seafloor to the oil deposit directly beneath, an easier proposition than taking the circuitous, slanting route from the wellhead to the oil, referred to as directional drilling, as some wells required.

Easier, perhaps, but definitely not easy.

The Macondo well would be far more than a hole in the ground. It would be an inverted skyscraper, a towering structure of steel and cement, telescoping downward ten times the length of the Empire State Building. That would be impressive enough, even if you ignored the most salient fact about what BP and its Transocean partners intended to do: The drilling would begin at the bottom of the ocean, where not a single worker could venture. Every bit of the construction of this hanging tower, the penetration of the rock, removal of debris, installation and sealing of the walls, had to be accomplished at the end of what was essentially a five-thousand-foot pole.

Even so, BP's plan for the well anticipated a drilling time of just seventy-seven days, at a cost of $96.1 million. The number of days and the number of dollars were inextricably linked. In business and industry, delays are always costly. But rarely is the cost of a delay quite as daunting as in the offshore oil business. Considering BP's costs—the nearly half-million-per-day rig lease, plus another half million a day, sometimes more, to pay for the rig's insatiable consumption of fuel, the daily helicopter flights bringing workers to the rig, drilling supplies, contractor services, food and catering—this meant that every day of drilling cost BP a minimum of a million dollars, every hour more than $40,000, every minute almost $700. That's day and night, seven days a week, as long as the rig was up and running.

The Marianas arrived at the Macondo site on October 6. Unlike the Horizon, which could do the work of drilling using its eight thrusters to remain virtually stationary above the well, the Marianas was attached to anchors set strategically around the target. Given the weight of the rig, and the huge potential costs of mooring failure, ordinary anchors—the kind tattooed on Popeye's forearm—wouldn't do. These were sixty-foot-high hollow steel cylinders with open bottoms that were driven deep into the silt of the ocean floor. Once sealed by the ocean bottom, air was pumped out the top, creating a powerful suction that kept the anchors firmly rooted. Now tethered securely in place above the target, the Marianas was ready. On October 7, the work began.

Although BP had officially set seventy-seven days as the projected length of the job, which would have meant that Macondo would be complete by Christmas Eve, they had dreams of going even faster. Well planners had drawn up a "target" of completing the well construction in fifty-two days, or by the end of November, which would give BP plenty to be thankful for—a savings of more than $20 million.

Nearly immediately, those aspirations took a hit.

Between the high pressures found deep in the earth, and the explosiveness and toxicity of the payload, poking a steel straw into an oil deposit that had slumbered peacefully for millions of years could turn ugly fast. The rig's massive blowout preventer—known as the BOP—was designed, in case of an extreme emergency, to shut down the well and sever its connection to the rig with the push of a button. It was a 53-foot-high, 325-ton, $15 million complex of valves installed on top of the wellhead at the seafloor. It controlled the flow of drilling fluids and cement into the well—and oil and gas out of it. The BOP's most powerful components were called "rams." If a well started to go out of control, the hydraulically driven devices—there were multiple sets arranged along the BOP—could clamp down on the drill pipe, closing off the hole until pressures were brought under control. If the situation went from bad to worse, one set of rams, the "blind shear rams," nicknamed "pinchers," were designed to cut straight through the drill pipe and sever the BOP stack into two parts, sealing the well below and allowing the upper part of the BOP to lift off the bottom and the rig to move away from danger.

Two separate control units, called the blue pod and the yellow pod, are attached to the top of the BOP and each can initiate any and all of the functions of the BOP when directed by signals that come down from the rig via communications and hydraulic cables. Even though it theoretically required only one pod to do the job, regulations—and common sense—dictated that both pods must remain fully operational at all times the rig was drilling to provide a safe margin of redundancy.

Less than a month after the Marianas began to drill, they hit an unexpected pocket of natural gas that got into the well and began to rise toward the surface. The buttons were pushed on the rig signaling the rams to close. The lower ram failed. The upper ram did close around the drill pipe, but its rubber seal was stripped away. Subsequent testing showed that the yellow pod had lost the ability to activate the shear rams, the last-resort pinchers.

Continuing drilling under those circumstances would have been illegal, and foolhardy. But the cost of fixing the problems was severe. The BOP had taken days to install. Now it had to be disconnected, hauled up on the rig, and repaired. Then installed all over again. Depending on how long repairs took, Transocean and BP were looking at a loss of between a couple of weeks and a couple of months.

The fifty-two-day target schedule was history, and now even the seventy-seven-day schedule was beginning to look unlikely.

Then the hurricane hit.

On the same day the managers on the Marianas reluctantly decided to pull their blowout preventer for repair, a small tropical depression entered the Caribbean. Within days it had strengthened to a hurricane. On November 5, Hurricane Ida passed over Nicaragua and headed north, straight toward Macondo. Rigs that could move under their own power, rigs like the Deepwater Horizon, scattered in retreat but managers of moored rigs like the Marianas had no choice but to pull all the crew off and leave the rig to ride the storm alone. The warm Gulf waters reinvigorated the disorganized Ida, which regained hurricane strength as it passed over the Marianas late November 8 into the next morning.

As the first helicopter approached after the storm had passed, the rig's senior managers could see that the Marianas had survived, but as the days passed, odd electrical glitches appeared. Electricians hustled to chase down the problems, but it soon became clear that the rig's wiring had been extensively damaged in the storm.

On Thanksgiving Day, the Marianas crew threw in the towel, unmoored the rig, and hitched her up to be towed in for repairs. A month had been lost. BP's dreams of a Christmas bounty were dead.

The holidays weren't looking much merrier for Transocean execs. On December 23, workers aboard their North Sea rig Sedco 711, who had sealed off a well they had just finished drilling for Shell Oil in the North Sea, heard a loud noise and looked up to see dark liquid shooting out of the well and spraying across the deck. A blowout. Shocked that a well they thought had been locked up tight could erupt, drillers fought to bring it back under control before the gas ignited and the rig exploded in a fireball. Everyone else mustered at the lifeboats and prepared to evacuate.

Fortunately, the BOP functioned in accordance with its design. An operator on the rig pressed the buttons on a control panel. The signal traveled down a cable from the rig to a control unit on the BOP, opening a valve and forcing hydraulic fluid into the ram system. The pistonlike devices clamped down on the well pipe and sealed it off. The flow from the well stopped, injuries were averted, and just three barrels of heavy drilling fluid spilled overboard.

But executives from Transocean were shaken. Their analysis determined that the blowout happened when a mechanical seal—designed to be closed at the time—had been mistakenly opened when the Sedco crew was trying to scrape debris out of the well.

They found that early warning signs of an impending surge from the well were missed because the drilling crew believed the well was complete and didn't see a blowout as a possibility. The incident could have easily ended much differently—in fire and death and an unstanched flow of oil into the ocean. They decided to rewrite the core well control procedures and hold conference calls to discuss lessons learned.

Senior rig managers filed into their conference rooms at the appointed hour, pulled up chairs at a long table with microphones spaced along it, and turned their eyes to the screen mounted above them as the call hooked up and the big bosses appeared in two dimensions to talk about this freak accident in the North Sea. The two most significant points the bosses wanted to make: There had been a failure to remain vigilant about pressure changes within the well, and a lack of clarity on how to control them. They emphasized that a sense of "complacency" at the end of a well could lead to disaster. They also noted that the blowout had cost them eleven days of work, more than $8 million, and a "significant loss of reputation."

The last thing anyone wanted, they all agreed, was for the same mistakes to lead to a far bigger catastrophe.

CHAPTER SEVEN

X MARKS THE SPOT

January 2007
Port of Baltimore

If anything had become clear from Dave Young's adventurous career at SUNY Maritime, it was this: He would not be happy with an assignment on some run-of-the-mill merchant ship. He needed something more stimulating, to both mind and body.

Dave was a classic high-energy personality. Even now that he was safely through to the other side of adolescence, a time for a man to begin the process of settling down, he never could grasp the concept of "sitting and relaxing." He was always doing *something*—and the riskier the better. He devoted ever more time and energy to building and racing those tiny, crazy-fast fiberglass boats he'd learned to build with his dad. He took them out on the Sound and pushed them until the water steamed in his wake and the boat leaped and slammed through the waves, shuddering on the edge of disintegration. Then after races he'd go out to all-night parties in the Hamptons, drinking with his friends and looking for women. He found quite a few, but only one stuck, a particularly beautiful and unusually understanding woman named Alyssa.

It was an ancient story. Flirtation turned into love, love turned into commitment, and a spirited young man found himself making a career and a family instead of another blender of lime daiquiris.

He abandoned the bar scene for domesticity, and a job behind the controls of a fiber-optic repair ship out of the Port of Baltimore. Charged with maintaining the thousands and thousands of miles of fiber-optic cables that crisscrossed the ocean and made the World Wide Web possible, the ship's demanding missions and technical complexity focused Dave's mind and, for a time at least, satisfied his cravings. The ship defied wind and tide to hover over an exact point on the ocean floor while malfunctioning cables were diagnosed and repaired, relying on the same kind of sophisticated thruster system that held oil rigs stationary over a well site. Every cable break was an emergency, minutes mattered, and the ship's crew had to function calmly while dealing with urgent commands from shore-based management.

Dave didn't think about it this way, but his experience on the cable ship was perfect on-the-job training for work on a self-propelled oil rig. Quite a few of his classmates from SUNY Maritime had discovered that mariners were sought after—and very well paid—in the offshore oil industry. The new semi-submersible rigs and drillships, now floating off Korean drydocks at an accelerating rate, all required a crew of certified mariners, even if the vessel spent nine-tenths of its life treading water. Not only was the pay superior to salaries on other commercial vessels, but the schedule—three weeks on, three weeks off—was more family-friendly than that of a merchant ship, which could head out to sea for six months at a stretch.

Dave's pals, convinced his skills and personality would make him a good fit for the unique world of exploratory drilling, urged him to apply for a rig job. But he resisted.

Despite the pay differential, a lot of mariners didn't want to work on oil rigs. For one thing, there was the "outhouse standing on a garbage scow" problem. As one captain given command of a rig put it, "What am I going to do when other captains break out pictures of their ships—show them *this* thing?"

And then there was . . . standing still. Especially for someone like Dave, someone with a passion for speeding over the water, it wasn't easy to contemplate a career where your "ship" could be mistaken for a "platform" and might stay hovering over the same oil well for months at a time. It was true that the fiber-optic work did require periods of hovering, but they were relatively brief interludes between long stretches of plying the ocean from one cable break to the next.

Sailors like to talk about how "salty" something is. It's simple mariner bravado but, roughly translated, a person or an object is salty to the degree it is linked to the great seafaring traditions reaching across the ages to the Phoenicians, and beyond. Sea beards are salty, black coffee with a pinch of salt is salty, and tattoos of naked mermaids are truly salty. And while plowing through high seas in a midnight gale at ridiculously high latitudes might qualify you as an old salt, spinning circles around a wellhead while the driller calls the shots would certainly not.

It wasn't that Dave was the kind who felt the need to brag to his peers about the shape of his ship or job description. It was just that for the moment he was happy doing what he was doing. But as his family grew—he and Alyssa had a girl and a boy now—he was tiring of the unpredictable needs of the cable ship, which was always on call. Its mariners had to respond like firefighters whenever a ship's anchor snagged and severed a cable or a repeater station resting on the seafloor sprang a leak. Except for Dave, the alarm bell was his cell phone and instead of leaving a warm firehouse, he was leaving his family. The need for fast response also meant he had to live in Baltimore, and Dave and Alyssa both yearned to get back closer to home. If he were on a rig, at the end of each work hitch the company would pay to fly him—or any employee considered a supervisor or technically skilled—to any airport in the lower forty-eight states, regardless of how out of the way, or how high the airfare. He and Alyssa could live anywhere they chose.

Even when he was on the rig, communication home would be vastly easier. The average merchant ship had only a single communal e-mail address. The captain printed out the e-mail messages sent to individuals, then put them on the bridge for retrieval, where anyone could, and often did, read them. Only one or two phones served the entire crew, and the five-dollars-a-minute satellite connection charge was deducted from monthly paychecks. On a rig, multiple phone lines were available free to crew members. There was also full Internet access, a feature very few companies running traditional ships can afford to provide. With the entire rig sharing the equivalent of one DSL or cable modem's worth of bandwidth, speeds frequently slowed to that of dial-up but were sufficient to send messages, pay bills, view photos of the kids, or stay up with the latest news and sports scores.

Between the allure of the perks and the mounting bills for diapers, formula, preschool tuition, and college savings plans, the higher salaries available on oil rigs—nearly twice that of merchant ships—began to speak to him. It took a year from when Dave's friends inside Transocean began to court him. In Dave's mind at least, the advantages of making the move finally outweighed the disadvantages of starting over in new territory.

He had been told that anyone simply e-mailing a résumé directly to Transocean was likely to be disappointed. You had to have someone inside to promote you, and he did. He sent his résumé to a couple of friends who were mates on a Transocean rig, the friends tweaked the résumé to emphasize the skills they knew Transocean was looking for—particularly his experience with dynamic positioning on the cable ship—and put it in front of their Houston-based rig manager. The rig manager got on the elevator in Transocean's high-rise office building in West Houston and hand-carried the application to the HR office.

A few weeks later, Dave was in Houston for a daylong physical and drug testing. A few weeks after that, he was headed to the Transocean finishing school.

Like all rig employees, before Dave began work he had to attend the company training center in Houma, Louisiana. There he and his fellow trainees spent two weeks sleeping in bunk beds in a prefab dormitory and sitting in classrooms learning how to administer CPR and recognize sexual harassment. When they got out of the classroom, the real fun began. In a workshop on personal survival, trainees donned their Gumby suits, the thick neoprene wet suits with hoods, booties, and mittens that made whoever wore one look like Eddie Murphy circa 1981, and more to the point, able to float and survive for hours even in frigid water. The most popular session, and arguably the most important, was basic firefighting. Instructors, typically retired firefighters, would ignite engines, pans of oil, and entire steel-entombed rooms, having the students work together to put the fires out. It was a class that all mariners, like Dave, had been through before, but training that bears repeating.

Then it was time for the final. Though officially called Helicopter Underwater Egress Training, its informal nickname, the Dunk Tank, better described the experience. Trainees climbed into a steel cage, strapped on seat belts, then slid down a ramp to the bottom of a pool. The objective: survive impact, calmly remove your seat belt, hold your breath while politely waiting for the person in front of you, then squeeze through one of the steel cage's window openings. Failure is painful. Scuba divers are prepositioned to assist those who get trapped. Occasionally someone panics, ingesting water with an unforgettable taste of death.

New hires who survived the two weeks of training sessions were handed a diploma of sorts. It looked like a driver's license but could be far more important—a "rig pass," an official passport into the offshore oil industry, which might as well have been a winning lottery ticket in a region where any jobs were scarce, and high-paid jobs otherwise nonexistent.

Dave's first posting was as a junior officer, a dynamic positioning trainee, on a drillship. From the beginning it became clear that his friends had been right. Dave's high-powered personality could be a challenge in many situations, but it meshed beautifully with life on a rig. It's as if people whose brains are built a certain way engage in dangerous hobbies or take risks not because they want to, but because they need to kick the brain into panic mode just to get enough brain stimulus to feel normal. On offshore rigs, stimulus overload *was* normal. Deep-sea drilling was inherently high-risk, high-reward, a continuously demanding series of complex operations involving multiple moving parts and quickly shifting objectives, not to mention the imperative to do everything *now*.

And just as in firehouses and police precincts, adrenaline junkies may have certain advantages on a rig, especially in emergencies. The flood of brain-stimulating chemicals released during times of danger may overwhelm some, but for those who crave it as a matter of equilibrium, risk may provoke a state of placidity that could make them the most functional people in the room during a disaster.

In any case, Dave thrived, and after a few months, he was promoted to second mate on a rig with a stellar reputation.

A blaring alarm tore into Dave Young's dream and prodded him awake. No hint of light rimmed the curtains. For a full minute, he couldn't say where he was. Then he remembered—a late flight to New Orleans, and a few hours of sleep at an airport hotel. He looked at the clock—5 a.m.—just enough time to dress and drag himself downstairs to catch the crew bus to the heliport.

Several dozen others were already there waiting, rig pass in hand, to check their flight order on the dispatcher's clipboard. There wasn't much suspense. The crew was listed by position, top down. The first names on the list had a seat on the early bird, and the last names could count on a day of waiting room magazines and Fox News droning in the background. While they sat there, they could wonder if they'd be tapped on the shoulder, handed a test tube, and escorted to the restroom to provide a sample for random drug and alcohol screening.

Much had changed since the good old days when most rigs had rollicking onboard taverns where off-duty workers could unwind. Rig workers could thank the captain of the *Exxon Valdez* for the abrupt end to that custom. Captain Joseph Hazelwood admitted to having "two or three" vodka drinks the evening his oil tanker drove onto a reef in a pristine Alaskan bay in March 1989, spilling ten million gallons of crude. Hazelwood was asleep in his cabin at the time of the grounding, and he was cleared of the charge of intoxication at his trial, but ever since the incident, drug and alcohol testing is an industry standard among U.S.-crewed ships, and very few now board a ship or rig under the influence. Almost to a man, the Deepwater Horizon's crew wouldn't even dream of trying.

Guys had such fear of the consequences that they often arrived for their hitch with foul morning mouth, having avoided toothpaste completely for fear of it reading as alcohol on their Breathalyzer test.

Some workers still found a way to smuggle something aboard for off-duty consumption, a caper planned and executed with the care and attention to detail of a prison break. More common were those who viewed rig time as enforced detox. When their hitch was over, if they wanted to, they could go on a weeklong bender. But they were damn sure to stop drinking soon enough to pass the test at the heliport.

Every trip to the rig via helicopter came with another test as well, a test of faith. If you asked rig workers to rank their greatest on-the-job fears, crane accidents, hurricanes, and even blowouts would no doubt rank behind the helicopter commute.

Every few years, a helicopter ferrying rig workers goes down, and in some years it happens more than once. The worst catastrophe came in 1986, when a Chinook helicopter carrying forty-seven passengers and crew from the shores of Scotland to a North Sea rig went down in a storm. Only two survived.

Even when the weather conditions are perfect, a small error in navigation can create large problems. In 2007, when Transocean's Discoverer Deep Seas finished up one well and moved a few miles away to begin another, the communication chain broke down and the incoming helicopter flight headed for the wrong spot. It arrived on location well past its point of no return, the point in the flight plan when there is no longer enough fuel to make it back to shore, and could not find the rig until it was almost too late.

When a helicopter does arrive, the danger is far from over. Two objects, one heaving up and down, rolling port and starboard, and the other jumping with each unexpected gust of wind, do not greet each other easily. Even after the craft touches down, a gust of wind can blow the long, narrow fuselage across the deck into personnel, steel bulkheads, or over the side into the ocean—a real-life dunk tank. As a result, more often than not, helicopter pilots do not shut down while refueling, but maintain reverse thrust with the rotors to hold the craft in position. The personnel disembark hunched over in fear of the rotors turning above their heads, grab their heavy luggage, and shuffle across the slippery deck. Until they descend stairs into the enclosed helicopter waiting room, they are not completely safe. In 2003, a pilot lifted off the deck of a Transocean rig in India and tripped over a net laid down to prevent her skids from sliding off the deck. As the chopper tilted over its blades dug into steel and shattered, launching steel fragments in every direction with such force that some were found embedded in the steel legs of the derrick more than one hundred feet away.

Of course, thousands of rig transport flights come off without a hitch, and Dave's first trip to the Deepwater Horizon was no different.

Now all he had to do was get acclimated with the motion beneath his feet. Ships, with their single V-shaped hull, tend to rock like a cradle. Rigs roll in circles like a cork, which can be unsettling even to the saltiest sailor. The good news is that waves can pass almost unnoticed above the submerged pontoons and beneath a rig's main deck. Even ten-foot waves have little effect. Rigs in the Gulf of Mexico can be so stable that some even have pool tables, which are playable more often than not.

Some coming off merchant ships found it difficult to adjust to the rig's chief asset, its phenomenal stability—the rig's ability to just *sit* there. As a junior officer, Dave's primary duty was standing a watch, choosing a path around ships, boats, and underwater obstacles, objects that might be invisible to the naked eye but visible on radar, or vice versa. It required both quick thinking in the moment and planning executed with mathematical precision and foresight. But on a rig that doesn't go anywhere 90 percent of the time, the job of standing watch on the ship's bridge boils down to long hours staring at computer screens and calling approaching ships, begging them not to hit you. It is boring. It definitely isn't salty.

But Dave was a quick study, and all the aspects of rig life soon became second nature to him. As his friends had supposed, his mechanical skills quickly made him a valued addition to the rig crew, and his personal qualities allowed him to float above the cultural divide that had proved so tough for other mariners to hurdle. For one thing, there was the lack of separation between officers and crew members. There were no fancy uniforms or rank insignia. Everyone referred to each other by first name. A mate's cabin was like a roughneck's, and there was no officers' mess—they all ate the same food. The culture shock was compounded for some because of offshore drilling's roots in the Gulf. It had always been primarily the province of a southern and largely working-class culture. For one thing, the lower-paid workers, who had no company travel allowance, couldn't afford to fly home every time they came back onshore for their weeks off. By necessity, they tended to live in or near the bordering states of Alabama, Mississippi, Louisiana, or (east) Texas. They all seemed to have nicknames like Big Country, Chickenhawk, Cornbread, Corn*fed*, and every rig had a Smokey—the guy who first caught the ship on fire. Off the rig, they tended to return home to houses at the end of country lanes where they could indulge their desire for acreage, even if they had to drive an hour for a bottle of milk. They had cows on the land, or chicken coops, or ATV hunting trails. They dipped snuff and owned shotguns.

It all could be a little alienating to someone who had grown up in suburban New York or the woods of Maine or on the beaches of California. To many of the mariners, the rigs could appear to be a redneck haven. But Dave took the ribbing often aimed at outsiders in stride, and he pulled his weight. He kept mum about his engineering degree and got his hands black with oil and grease alongside his crew. He discovered that if you were willing to accept the culture and the occasional comments about where you grew up, you could advance—quickly. If you came in with attitude or took offense to the jokes, well . . . that didn't get you anywhere. The most important attribute on an oil rig is the ability to work and get along.

Work is the stuff and substance of rig life. Everything is designed to keep the rig working every minute of every day. There are two shifts, each twelve hours a day, seven days a week, with very few breaks. If there's an ongoing operation on the rig floor, lunch waits. Food is wrapped in tinfoil and put away, devoured when there's time—even if it's only five minutes. There are no unions offshore. Only company policy and a supervisor's goodwill stand between men and exhaustion. If the company wants drillers to work sixteen hours straight, they can make them.

The normal compensation for the grueling schedule is the promise of three weeks off at the end of every three-week hitch for supervisors and technicians (or two weeks on, two weeks off for lower positions). And the money: Drilling hands, kids right out of high school, can make $40,000 to $50,000 a year. As third mate,

Dave was bringing home close to $100,000 a year. The captain earned up to $200,000 and the OIM even more.

The money was good enough to make rig life bearable to most, and even enjoyable to many. Weeks on the rig were a time to be away from the complications of shore life. They were with friends, doing challenging work that brought tangible results. They worked hard, and when the long shifts were over, they ate dinner, read a book, watched a movie, or went to bed. Every evening, you could find a handful of romantics at the deck rail, watching the sun set. For every nature lover, however, there were half a dozen others glued to video consoles, playing Street Fighter. Some of the younger hands had the energy to work out in the small "gym," which was just a room with a weight set, two treadmills, and a StairMaster. Others played poker, friendly games mostly, though some might leave the table a few hundred bucks poorer or richer.

Others, like Dave, spent most of their nonworking, still-conscious time on laptops, chatting with wives or girlfriends, working toward online degrees, starting personal websites.

Or watching television. The rigs had satellite TV capable of receiving twenty channels at a time—making inevitable the ongoing battles over *which* twenty. The perennial winners: Fox News (always), the Weather Channel, Country Music Television, the Rodeo Channel, Outdoor Life, and anything involving antlers and largemouth bass. When the big bosses came in from town, CNBC was switched on. Most rigs used to have adult channels, but as more women began to arrive, the racier TV fare was purged. Instead, contraband porn DVDs were traded around the rig like baseball cards.

The ten-by-twelve-foot crew cabins, most of which had a DVD player and a seventeen-inch TV, were identical. Most were shared by two crew members who worked opposite hours—meaning they were almost never in the room at the same time. Some low-level crew were bunked four to a room. Only the captain and the OIM had a room to themselves.

Legend has it that the captain of Transocean's rig Discoverer Deep Seas, to better demonstrate his elevated status, requisitioned a monster forty-two-inch flat-screen TV that barely fit in his tiny room. Word of the purchase order, approved by the rig manager onshore, leaked out. As soon as the hitch was up, some crew members visited a local florist to order the biggest, gaudiest flower display in the shop, complete with a teddy bear centerpiece and balloons. They sent it to the rig manager with a note that said, "Thank you for the flat-screen TV!"

The TV order was promptly canceled.

Ribbing and one-upmanship were common elements of rig life, but outright fights were rare. The lack of booze and drugs, and the fear of doing anything to jeopardize the almighty rig pass, tended to keep things civil.

Rig populations are still overwhelmingly white and male. Transocean and other companies have made some advance in diversity in recent years, but the largest incidence of women and minorities still occur in the catering staff. Given the relatively low salaries and long hours, many of the catering positions are filled by people who come from impoverished backgrounds and have few other options. On some rigs, the caterers, hired and managed by a subcontractor, remain permanently on the edges of rig life, never really thought of as part of the crew. But some crews make a point to include the cooks and waiters and janitors, inviting them to weekly safety meetings and drills and other communal events. You can always tell when you're on one of those rigs by the superior quality of the food and service.

A lot of people knew the story about the rig with an openly

gay baker—an out homosexual is the rarest of rig rarities. This baker was a talent, and his specialty was elaborate birthday cakes, which he delivered in person dressed as Marilyn Monroe to sing "Happy Birthday" to the embarrassed recipient. The crew all but marched on the rig manager with electric torches and dinner forks demanding that the baker get bounced back to shore. But the baker stayed on. A couple of years later, word got around that the catering company was finally getting ready to let the baker go. By then, the quality of his cakes had trumped the crew's homophobia. Or maybe they had learned to enjoy his singing voice. Some of the men who had screamed loudest for his head successfully petitioned management to keep him on.

The meals on the rig are all-you-can-eat, and almost every meal features something fried in fat or butter with high rations of salt and sugar. Any attempt to provide healthier fare is met with fierce resistance. The rare healthy eaters among the crew tend to load up their suitcases with cans of tuna and good coffee, and wind up eating a *lot* of cereal. Fresh food like milk, eggs, vegetables, or fruit is always welcomed by those wishing to avoid the lethargic aftermath of fried okra, but if there is bad weather or a logistical problem and a new grocery box doesn't come out on the supply boat for a week, these items quickly disappear.

The rest of the food never does. On the Horizon, the break room cabinets contained a bottomless supply of cans of Beenee Weenee and Vienna sausages, bread, peanut butter and jelly. Every four hours, a more ambitious spread appeared there—sandwiches, pizza, cookies—free for the taking. On Sundays, the crew could look forward to the rig barbecue. The deck would sprout folding tables and chairs and half barrels loaded with ice and stuffed with cans of soda and, if BP was feeling generous, bottles of nonalcoholic beer. The barbecue grill itself was the pride of the rig, where it was kept in a place of honor on the deck. The Horizon had its Korean trophy grill, but some rigs were known to spend as much as forty thousand dollars on machines so complex they looked like they might be able to circumnavigate the moon with smoking attachments you'd need an air winch just to lift onto the deck. But they admirably fulfilled a grill's basic function, turning out well-charred steaks, lobsters, and twice a year, big buckets of boiled crawfish shipped in for the occasion.

The whole affair looked like something you'd find at the end of a dirt road in the Mississippi outback, except no one is smoking. Most tobacco on the rig is chewed or dipped. Spitting on the decks is strictly prohibited, but anywhere off the leeward side of the rig and trash cans are fair game. On an oil rig, you don't ever want to have to go digging through the garbage can. Those who dip place a paper cup lined with a crumpled paper towel in their shirt pocket, allowing them to spit with just a tilt of the head. By the time those cups are tossed away, they are pretty full.

Smoking is limited to two areas outside on the perimeter of the lower deck, each equipped with a couple of benches, a butt can and an electric lighter mounted on the wall. There's also a red light—a signal from the computer system mandating that all smokes be extinguished, immediately. When it flashes, crew members tend to pay attention: The light is prompted by the flammable-gas detectors, and is a reminder that beneath them is enough explosive gas to blow the rig out of the water.

Danger is a constant presence on an oil rig and something that few can erase from their minds.

But blowouts and mass explosions rarely top the list of worries. Most companies, including BP and Transocean, spend immense time and effort attempting to prevent the most common hazards: loss of fingers, back injuries, minor chemical burns, slips, trips, and falls. Transocean had an official "vision" for safety, and it was repeated with near religious fervor: "Our operations will be conducted in an incident-free workplace—all the time, everywhere."

To walk into a Transocean workspace is to be accosted with warnings. Every bump in the floor over which you might trip is painted a bumblebee pattern of alternating diagonal yellow and black stripes brushed on the steel deck. Affixed to the walls, marking every conceivable hazard, are color-coded signs—green for safety, red for fire or explosion, yellow for danger—embossed with silhouettes of figures acting out whatever it is you should not, under any circumstances, be doing. A decade ago, Transocean, prompted by BP, hired a company called Seward Signs to survey an aging Transocean rig. Seward representatives, wearing logo-embroidered, bright white coveralls that appeared to cost as much as any designer apparel, were flown from the United Kingdom to spend a week crawling through every compartment of the rig looking for potential hazards they could mitigate by installing a sign. Weeks later a box arrived via FedEx containing bound books, each over two hundred pages in length, listing their recommendations.

Even when instructed to pare down the list to the bare essentials, the quote came in at more than one hundred thousand dollars.

And signs were just the beginning of the safety spending. Lifeboats, each costing the amount of an average American house, topped the list, but in aggregate the costs of preventing minor injuries was much higher. Teams of contractors descended on every newly built vessel to push safety videos, manuals, and contraptions of all types aimed at ensuring a safe work environment. They found a receptive market in Transocean executives, who competed to champion innovations in shipboard safety before they filled out their end-of-year self-evaluation forms, which had a place to list them. The emphasis on safety had resulted in some excellent ideas, such as positioning automatic defibrillators on the rig floor and in the engine room, but not all the ideas proved so practical. One company scoured public industry incident reports and found minor injuries related to the use of sledgehammers. They devised an alternative tool, a pole that would be centered on a stuck bolt with two handgrips, one to hold the pole steady, and the other to hoist up a sliding weight. The rig hand straddled the device and proceeded to pull up on the weight, a physical task accompanied with a grunt, then let it drop down the pole. Of course the device didn't loosen bolts very well, but it was nevertheless used because the repeated action—pelvic thrust, grunt, pelvic thrust, grunt—always managed to illicit an adolescent response from coworkers.

Another safety device sold to the rigs was called a "backscratcher," a steel cage affixed to vertical ladders to prevent a climber from falling backward off the ladder. But that wasn't safe enough. Transocean managers ordered containers full of wires, harnesses, carabiners, fall arrestors, and mounting gear, along with technicians who would bolt failsafe devices to any ladder over six feet high. Rig hands were then required to don five-point harnesses and clip into the failsafes. One employee became famous for voicing the prevailing sentiment: "I was never afraid of climbing a ladder until I came to work for Transocean."

Though there were always some who griped about what sometimes seemed like overkill in safety initiatives, most rig workers—who, after all, had their fingers, toes, limbs, skull, and butts on the line—wholeheartedly supported the safety programs. Occasionally a little too wholeheartedly.

In 2009 Transocean asked each rig to produce its own footage of work situations that created a hand injury risk, for inclusion in a company-wide safety video. The idea was to get jaded rig crews to pay attention to these usually boring films by featuring real co-workers instead of hiring actors. One rig under BP contract—the Development Driller II—got the message, marked urgent, to send in video ASAP. The medic grabbed a willing crew member and, camcorder in hand, went down to the rig's machine shop. His instruction to the crew member was to put his hand near the blade of the device to demonstrate the dangers of improper hand placement. They turned on the camcorder, turned on the saw, and proceeded to capture a more gruesome scene than the managers had asked for.

For Transocean execs, even the company's cornerstone of safety—a mandatory time-out before even the simplest procedures to take a moment to assess potential dangers, called THINK—didn't guarantee the level of safety awareness they were striving for. So they created another program with yet another acronym whose initials were a mystery to almost everyone. It was called START (See, Think, Act, Reinforce, Track, as it happens) and required all rig workers to become impromptu safety control managers. At least once a day, employees were required to circulate the deck with a pocketful of index cards, each slightly longer than a dollar bill and printed with a checklist of potential hazards. When they saw safety principles being violated, they were supposed to check the corresponding hazard on the START card. Was someone missing a hard hat? Check. Safety glasses? Check. Was heavy material being moved without regard for possible pinch points or crush points? Check. Were there any slippery surfaces in a work area? And so on.

No minimum number of cards was required, but the common

wisdom on the rig was that "a START card a day keeps the rig manager away."

Every check mark for major or minor violations of safety rules, every missing earplug or tripping hazard, was logged in a giant computer database onshore. The numbers were tracked—the amount of cards submitted by each division, rig, and even each individual employee could be spit out in an instant.

Often the data was used to make new safety policies. One year too many cards came in about missing hard hats, so the managers bought chin straps to keep the hard hats from blowing off in the wind. Of course most of the missing hard hats had nothing to do with wind, but that wasn't the point. The point was that Transocean was tracking the potential for injury and actively working to correct problems.

The focus on safety expressed itself most immediately in the pre-job meetings—without which no procedure on the rig, however routine, can begin. All work is shut down and relevant personnel are gathered on the rig floor to review the procedure. The senior supervisors each take a turn saying things like, "I know you've all done it a million times, but you can't let your guard down. It's a big job. This is heavy machinery, heavy equipment, things can turn ugly, so keep your head on a swivel." Some of the southerners—thoroughly used to the idea that everyone within hearing was male—might try for something a little more colorful, along the lines of "Men, make sure you don't put your hands anywhere you wouldn't put your dick . . ." Then all the other supervisors would pretty much repeat the same message, beginning, "As Jimmy says . . ."

The men made eye contact and nodded, but most were just waiting to get on with the job. They knew the drill, knew they had to appear as if they were hearing all this for the first time, even as they were engaged in a raging internal debate over whether the terrorists took over the skyscraper in *Die Hard* or *Die Harder*. It wasn't that they didn't care about safety. But they were like frequent flyers who had to listen to the attendant's spiel about where to find the flotation cushions for the eight millionth time. They knew, they knew.

CHAPTER EIGHT

THE FLOOD

May 2008

Gulf of Mexico

One day in the spring of 2008, Dave Young was onshore, dry and comfortable in his Connecticut home, when he decided to check in with the rig. One of the watch crew, a dynamic positioning officer, answered the phone on the bridge. Dave could immediately hear the strain in his voice.

"Something bad is happening right now, right?" Dave asked.

"Yeah," the DPO replied.

"How bad?"

"I've got to go," he said, and then hung up.

The crisis began, as crises so often do, with plumbing.

By the time the Horizon had been on station in the Gulf for seven years, it had drilled more than twenty wells. Middle-aged now, it was no longer the newest, most advanced rig in the fleet. The maintenance requirements grew longer with every shift. The Horizon was showing its age.

And the staff had shrunk.

Doug Brown hated to see anyone go, bound as he was with his crew. But his frustrations were growing less from personal concerns—the sort that kept Jack Parento's coffee cup in the engine control room seven years after his heart attack—than from professional ones.

When the Horizon left Korea, his department had a chief engineer, a first engineer, two second engineers, two third engineers, and four motormen. A few years in, Doug saw his staff begin to thin. Transocean removed a motorman and a third engineer, followed nine months later by the first engineer. At first, Brown could understand. In the rig's early years, when all the equipment was new, he had to admit some of his people were just sitting around. But as the years went by, he found that things started to break down more often.

The workload increased dramatically.

Doug still loved his job. Each time he began his hitch he'd go down into the engine rooms to check up on his babies, the six gargantuan diesels, and see how they'd fared in his absence. "Did you miss me?" he'd ask. It wasn't just the engines he was concerned about. Doug was the shepherd, and his flock consisted of all the moving parts of the rig not connected with the drilling machinery. If air-conditioning went down in the accommodations, Doug and his crew repaired it. If the fresh water started coming out salty, they tackled the desalination system. A thruster wasn't giving full output? Doug got into the innards and set it to rights. But increasingly, he couldn't give his charges all the attention they needed. He began to feel that preventive maintenance was going unattended. Whenever Doug expressed his concern that he was oversubscribed and understaffed, he'd get no satisfaction from the response: "I'll look into it," he'd hear, or "Town is working on the problem."

"Town" of course, meant the corporate headquarters in Houston, which too often seemed to be located on another planet. In any case, Doug got no relief.

Many of those who found themselves dealing with any but the most urgent problems, ones with the visual evidence of burnt parts, water spraying high and low or when the drill bit stopped spinning, quickly became frustrated with time-consuming paperwork. The administrative hurdling began with detailed documentation of the problem in EMPAC, the rig's computerized records management database programmed in the late 1990s (and by all appearances never really updated) for warehouse managers and shoehorned, it seemed, to fit Transocean's needs offshore. They found the system counterintuitive, with cumbersome drop-down menus, complex coding, and archaic search capability that seemed designed to make it as difficult as possible to find what they needed. The request was followed by a wait for authorization to order parts. Then, even if the parts were right there in the rig's warehouse, yet another EMPAC work order was required to actually get them issued. If a spare part was not aboard, it could take weeks, sometimes months, for Transocean's buyers to get it there.

With parts in hand, a final risk analysis had to be written and signed, lock-out notices posted warning people away from out-of-service equipment, and permits to work authorized by multiple parties before a repair could proceed.

The combination of shrinking staff and bureaucratic processes added up to a growing list of deferred maintenance needs, and there's more than one way that can make trouble. Small problems can become big ones. Things can break at just the wrong time. Or an overwhelmed maintenance crew can begin making mistakes.

Sometimes the mistakes go unnoticed. But sometimes they can cascade, as they did in May 2008.

The Horizon's constant exposure to salt water combined with the unyielding element of time had degraded miles of complex piping that carried the water used for everything from ballast to engine coolant. In the course of drilling more than twenty wells in seven years of service without ever touching shore, the Horizon had reached the point where the obvious solution would be a complete replumbing. This would have required the rig to be drydocked at huge expense. Instead, it was decided that the rig mechanics would begin a patchwork replacement of the most corroded steel.

One section of pipe identified for replacement was a spool piece on the discharge side of a saltwater pump in the forward starboard support column, seventy-five feet below the waterline. The pipe carried seawater to auxiliary systems like the freshwater maker and the thruster cooling system.

The engineer who removed it was a maritime academy graduate with an engineering degree and experience interpreting the spaghetti-like piping diagrams. He was required by job description to know every valve, pipe, and pump function. But he was also pressed by the huge backload of maintenance and the workload that came from being undermanned. Skipping the paperwork that would have made sure everyone knew of the ongoing repair, he simply called the bridge to let the DPO know what he was doing, then removed the corroded pipe.

When his shift changed, he turned the job over to his partner and went to bed. Instead of completing the repair, the relief engineer went on to other projects.

Meanwhile, the watch team also changed shifts. On the afternoon of May 26, 2008, the new DPO noticed that the rig's ballast was off-kilter. He opened a series of valves to correct the rig's balance, including the out-of-service valve the mechanic had failed to lock out. Seawater—which weighs about as much per volume as steel—gushed out from where the corroded pipe had been removed, quickly flooding the pump room.

A bilge alarm sounded. Chief mate Marcel Muise and an assistant DPO went to investigate. They hadn't gotten far when another alarm sounded for the Number 2 thruster compartment. The flooding was spreading. Marcel called the bridge and ordered a general alarm.

By 8 p.m., seventy-seven crew members had been transferred to a nearby workboat to wait for the danger to pass.

Before the personnel who remained on the rig could attempt to fix the problem, they would have to make sure they understood what had caused the flooding and why it was spreading. An emergency team would have to descend, far below the waterline, to the flooded area in the support column.

It was just about then that the phone rang on the bridge. It was Dave Young, checking in.

The DPO couldn't stay on the phone even for a second. He needed to focus all his attention on the rig's vessel management computers, which, among many other things, controlled the remotely operated valves that directed water within the rig's ballast system. Theoretically, the bridge team could, with a few clicks of the mouse and the opening and closing of valves from the emergency ballast panel, reverse the plumbing and pump the flooded chamber dry. But the flood had fried the sensors that indicated which valves were opened and which were closed. They were flying blind, which meant they would have to be extremely careful. The mistaken opening or closing of one wrong valve could turn a tense situation into a disaster.

To rectify the problem, they would have to find the exact positioning of the relevant valves so they could realign them to pump the flooded chamber dry. That would have been one thing if the valve sensor system had been operating. Now that it had been compromised, they'd have to do it the old-fashioned way.

Marcel darted back down the aft column to work his way through each compartment of the starboard pontoon forward to the pump room where the section of pipe had been removed. The heavy watertight doors between one compartment and the next slowed his progress. Finally he approached the second-to-last door before the flooded compartment and throttled the hydraulic lever to open it. It was a mistake. Marcel had made an assumption that the water hadn't spread this far. The fallacy of that assumption poured down on him. He had to close the door, and fast, to stop the compartment from filling with water, but the door throttle was already consumed by the flood. Instinctively he pushed his body into the torrent and grabbed the lever. The powerful door pushed back against the surge and closed tight, but the pressure was so great, water was still spraying through the door gasket. Lightbulbs shattered and the power went down. Marcel's radio no longer could transmit in the absence of the VHF repeater signal. Wet and cold in the dark, he knew he needed to think clearly. If they didn't pump the water out, and fast, the slow spread could become a rampaging flood.

He hurried back to the bridge, where the rig's senior leaders were also having difficulties. The DPOs pored though complex line diagrams of the plumbing, waiting for the company's engineering department ashore to assist them in finding a solution. Assistance never came.

Despite cabinets and computer servers overflowing with technical drawings, no one in Houston could seem to locate the rig's piping diagram at this late hour. Marcel had tried to e-mail the needed diagrams as soon as the flooding began, but he discovered that the Internet connection had been shut down by the OIM. Marcel thought he knew why: No OIM wanted to see his rig on CNN.

Marcel huddled with the watch crew to figure out the needed alignment of valves and pumps on their own. After working feverishly, they thought they had the solution. If they had gotten it wrong, instead of pumping water out, they could be pumping it in. Marcel, the chief mechanic and the toolpusher, went back down the column to make sure that didn't happen. The watertight doors to the flooded compartments were still holding, but water was gushing from a hole behind a control panel in the adjoining wall. It was a small hole, so the water flowing through it was negligible. But Marcel realized that since he couldn't see inside the compartment to know if the water was subsiding, this hole would provide the confirmation he needed that the plan was working.

The pumps were engaged, but the water kept spurting through the hole. A valve they thought had been closed must have been open. But which one? It took several tense hours for them to manually track and test all the possibilities, but finally they found the open valve and shut it. They fired up the pumps again. This time the telltale spout of water slowly lost force, then stopped. They were in the clear.

It could have gone differently. In a vessel like the Horizon, so big and complex, even a small problem could lead to bigger problems and threaten to plunge out of control. A hasty repair had led to a miscommunication that resulted in a flood. The flood shorted out sensors critical to diagnosing the problem and pushing the water back out of the rig. If the cascade of bad luck and bad consequences had continued, it conceivably could have all ended in a half-billion-dollar rig capsizing and sinking to the bottom.

By 2 a.m., though, the flooded compartments had been pumped dry, ballast had been rearranged to correct the list, and all personnel were back on the rig. No injuries or pollution resulted from the incident.

The following day, a Transocean team led by Buddy Trahan, an experienced executive who had worked at almost every possible rig job, arrived to help the Horizon's people get the Horizon cleaned up and ready to restart operations. It would take several months to repair all the bilge sensors, valve indicators, and pumps, but thanks to the Horizon's built-in redundancy of functions, the crew was back drilling ninety-six hours after the incident. A collective pride glowed from the drill floor to the bridge. Nature had pressed hard against the rig's inner walls but the crew had averted disaster using teamwork and experience, with no assistance from shore.

The investigations everyone feared, the ones that could shut down operations, were canceled. Neither the Coast Guard nor the federal Minerals Management Service (MMS) decided to pursue the incident with investigations whose focus could have widened to include the Horizon's other urgent maintenance needs. The readiness of the MMS to close the books on the incident was to some degree understandable. The federal agency charged with managing the nation's oil, natural gas, and mineral resources was called to investigate thousands of serious oil-related incidents each year. In a recent three-year period, the incidents included 30 worker deaths, 1,298 injuries, 514 fires, and 23 blowouts that left wells out of control.

For all of this they had just sixty inspectors.

The relief lasted only until the next morning, when another helicopter brought a joint BP/Transocean investigation team out to the rig and the expected praise turned into admonishment. Investigators made the conference room over into a war room, filled binders with photocopied permit-to-work documents and interviewed key personnel. Despite the considerable energy expended and some harsh conclusions, the investigation seemed to miss some important lessons, like the need for better engineering support from Town, the strain that maintaining an aging vessel put on a shrinking technical crew, the paperwork blindness caused by a burdensome bureaucracy.

The investigators did not leave empty-handed. After making their way through the trail of paper, they faulted the original mechanic for failing to file the warning notices. He was terminated and eyes turned to the rest of the crew, finally settling on the Deepwater Horizon's captain.

The easiest questions in the Coast Guard's merchant marine officer licensing examinations always go something like this: "If a seaman slips on a wet deck and gets a concussion, who is responsible?" Or it could be about a faulty valve deep in the bowels of the ship's engine room that opens to leak oil in a bay, or contraband found in a shipping container sealed long before it ever arrived on deck, or a dispute over a toothbrush. Or a flood.

There's only ever one correct answer: the captain. The captain is responsible, always. Within weeks, Transocean would be looking for someone new to skipper the Deepwater Horizon.

CHAPTER NINE

A CAPTAIN'S COLORS

June 2008

Baltimore County, Maryland

Curt Kuchta had that college quarterback look—six foot three, square jaw, dark complexion, bright smile—the easy smile of someone confident that he was going to be liked. Curt graduated from the Maine Maritime Academy in 1998 and found his way to offshore oil exploration and Transocean the same way Dave Young had, through the web of contacts he'd made while a student at the academy. In fact, he'd married the sister of one of Dave's SUNY Maritime dorm mates.

Curt loved his work and was good at it. The photo he chose for his Facebook profile showed him happily dangling from a harness, doing maintenance on the side of a lifeboat. He loved the rig lifestyle, which allowed him to live in a pleasant Baltimore suburb even though his commute was 1,200 miles each way. He had three weeks out of every six to focus on enjoying life with his wife and young children, whom he indulged with trips to amusement parks and train yards and an endless variety of games, which he enjoyed almost as much as they did. Curt also liked playing with his man toys, afforded by Transocean's generous pay. He went fishing on his deep-V Grady-White ocean fishing boat powered by a 225-horsepower Mercury, skidded in the mud in his off-road truck, or plopped his son on his lap and ran circles around his yard on his John Deere riding mower. Once a year he got together with his old maritime academy friends for an annual pig-out and pigskin event that commemorated the life of his wife's brother, who had died two years after graduating from SUNY Maritime. They drank beer, played touch football, and told real and embellished sea stories—all of which Curt excelled at.

Aboard a Transocean rig, you could tell a lot about a person by his appearance—especially in the Gulf, where fiercely independent American southerners had refused to accept the Transocean standard of coveralls for everyone. The workers complained to the managers and the managers stood up for them at headquarters by resisting the coverall mandate, and even delaying for years changing their uniform color from ocean blue to red, despite the obvious safety reasons. But it took Transocean three years to win even that battle, and the company gave up entirely on mandating coveralls. The corporate compromise allowed three types of dress; the red coveralls, jeans with a red button-up shirt, or bib overalls, of a type worn by farmers, with a red shirt underneath. These inevitably became cultural markers. The bibs were selected by those who wanted to emphasize their country roots, while the jeans were the choice of supervisors who didn't mind referencing a subtle class distinction.

Curt chose the jeans and added a belt with a large brass buckle embossed with the Horizon's image and a Transocean "T" logo. Since first entering Calvert Hall College High School—a private Catholic school—then going on to Maine Maritime, his uniforms had always set him slightly apart from general society, which was fine with Curt. He took pride in his appearance.

The crew had the option of two styles of hard hat: one with a full brim circling the head and another with a visor in front. When Curt first joined Transocean, this had been a false choice: the good old boys who controlled some of the company's older rigs issued hard hats with visors to anyone considered to be a Yankee, which in their strict accounting meant anyone living much farther north than Interstate 10. The insularity had largely faded away by the time Curt joined the Deepwater Horizon, but it still took a lot of confidence for a Yankee like him to choose a full brim. Curt had a lot of confidence.

But along with confidence he had some striking commonalities with his southern colleagues, from his casual but courtly charm to his staunchly conservative political views. Like a great majority of those he worked with, nearly all except a smattering of "bleeding hearts," he preferred Fox News to CNN and Limbaugh to NPR.

The final clues to his personality he wore on his forehead. Literally.

Transocean was so focused on making things run smoothly, top executives decided it might help if every employee wore his personality on his hard hat for all his coworkers to see. To accomplish that, all new hires, at all levels of the company, were administered an hourlong diagnostic test in which they were asked mind-twisting questions like "Do you think you are timid or courageous?" or "Are you generally introverted or extroverted?"

The theory is that the answers reveal four types of personalities, and that each individual is a combination of two of the four, with one personality type dominant and another a secondary influence. For easy reference, each type is associated with a color that must be displayed, via red, blue, green, or yellow stickers, on everything from hard hats to office doors.

Green is the color of a thinker, a perfectionist, someone who is precise but also critical, picky, argumentative, and slow to make decisions. Blue is the color of the feeler, someone who tends to be dependable, agreeable, supportive, and calm, but also reserved, awkward, possessive, and insecure.

Curt's dominant color was yellow—the color for socializers, who, according to the analysis at least, tend to be enthusiastic, optimistic, talkative, persuasive on the upside, disorganized, undisciplined, and at times overconfident on the downside. Curt's secondary color, red, was common among upper-level supervisors, signifying high ego strength, strong will, a desire for change, but also someone who is pushy, impatient, domineering.

There was one thing about Curt that no one needed personality tests to see: He was a hard worker, an important trait at Transocean, where there was little tolerance for anyone with the luxury of too much "ass time." Between his industriousness and his native likability, Curt was in the right place at the right time.

In the past, it had taken a couple of decades to rise through the ranks to make captain—the pinnacle for a merchant mariner, tantamount to membership in an exclusive club that all but guaranteed lucrative employment for life. But as the search for offshore oil expanded rapidly after 2001, promising young officers could rise quickly, and Curt was well on his way. He'd gotten hired on an R&B rig as third mate right after graduation in 1998. In 2004, he made chief mate of the Horizon, by then a Transocean asset. He'd always received high marks on his evaluation reports, and by 2008, just ten years into his career, Curt's hopes for making captain got a huge boost. A Transocean friend and colleague—another chief mate on a nearby rig—walked into the HR office at Transocean headquarters in Houston and got a look at something called the "ascension list," the short list of chief mates in line to be promoted when a suitable captain's spot opened up. Curt's name was there, third from the top.

When the company announced the building of new drillships in Korea, however, Curt had an opportunity to take the chief mate's spot on the new rig. He took it, even at the risk of losing his place in line on the ascension list. He was always up for an adventure, and going off to Korea as part of the all-star crew that would help oversee the construction of the latest technological marvel definitely qualified. So he put his captain ambitions on hold and accepted the chief mate job.

But he never made it off the drydock in Korea. Back in the Gulf, the Deepwater Horizon's pontoon inadvertently filled with water just two months after Curt's departure. Since Curt's history on the Deepwater Horizon counted for even more than a slightly higher spot on the ascension list, Curt leapfrogged over the names ahead of him and got the call. He was just thirty-two, and he had made captain of a half-billion-dollar rig.

The fact that—except for the belt buckle—Curt's work uniform made him look more like a building code inspector than a ship's captain was worth contemplating. The traditional image of a captain—an aloof, imperious, and unquestionable authority—formed in a time when ships in the middle of the ocean might as well have been alone in the world. Things were very different on an oil rig. For one thing, in daily phone and Skype conferences, both Curt's bosses at Transocean and the clients at BP tried to steer things from "the beach," often crossing into micromanagement. Even on the rig itself, the captain was officially subservient to the offshore installation manager except when the rig was moving from one well to another, which was less than 10 percent of the time. And both were beholden to the "company men," BP project managers who lived aboard the Horizon.

BP kept two company men on board the Horizon at all times. The Transocean personnel managed all the rig's activities, from drilling to maintenance, but the overall strategy of the project, and the well design itself, was all controlled by BP engineers and contractors with Houston offices, all of whom reported to a shore-based well team leader. The company men, who were called well *site* leaders, made sure the concerns of the well team leader were attended to on the rig.

The company men held a special place in the chain of command, an off-the-books dotted line on the organizational chart. While not exactly able to issue orders, they were in the position of homeowners haunting a construction site to ensure the contractors satisfied all their expectations. Few questioned their authority. Further, they got the best rooms on the rig, their choice of channels on the TV system, and if their laundry came back wet or stained, everyone would hear about it and the problem would be quickly rectified. Besides having the power to make the laundryman's life miserable it was their job to keep the train on track and on schedule.

According to Transocean policy, the OIM had the final say on all decisions, but in practice, the OIM's only recourse to an order from a company man was to stop the job and send the complaint to his immediate superior, the rig manager in Houston. This naturally caused his phone to ring incessantly with questions from the beach . . . something most OIMs strived to avoid.

Some company men took this power in stride and became one of the crew and an advocate for the crew's interests and the rig's safety during meetings with shore. But the majority of company men justified their informal title. If BP wanted the crew to jump, the company man made sure they jumped; if the shore wanted to give the guys a pat on the back, he provided the service. And if they wanted to pick up the pace he would start pushing.

Company men weren't out-and-out adversaries, but they did tend to maintain their distance from the Transocean personnel. They ate at the "BP table" in the galley, set apart from the crew, and they only communicated with the OIM and the toolpushers.

Even the captain was usually beneath their interest.

All of that was fine with Curt. He was spared the bother of daily client requests, and both his OIMs were veterans of the position who made it clear they respected Curt and the way he ran the "ship" parts of their rig.

There was no denying that his was an ambiguous position. Even in the eyes of his own crew, a captain's status was vague. All that most really knew about Curt's job was that his was the voice coming over the PA, the man in charge during Sunday emergency drills.

So while the title sounded grand, it was a good idea for rig captains to appreciate the reality of their situation. Those who tried to adopt the trappings of command only managed to provoke derision. Any captain who insisted on being addressed by rank got very little respect. Successful captains learned to be self-deprecatory, to make light of their own situation by saying things like " 'captain' is just for the courtroom" or "I'm only the bus driver." If you had to tell people you were in authority, you weren't. A captain's leadership should speak for itself. Because sooner or later, it would have to.

Dave Young's career at Transocean was elegant testimony to how fast a young officer could climb the ranks. In less than three years, he went from being a trainee to chief mate on the Deepwater Horizon, promoted in early 2009, less than a year after Curt became captain.

Making chief mate was a big step up. Chief mates are considered senior officers and the job meant leaving the tedious hours of standing watch to others. Dave was liberated to roam the deck. In fact, his job required it. One of his duties was ensuring the rig's stability. That meant tracking and stowing all heavy equipment so that no imbalances were created and maximum capacity was never exceeded. To do this, he needed to enter the precise location and weight of each item into the rig's stability computer. For Dave, with his degree in naval architecture, picking off the precise location of a piece of equipment from the auto-CAD drawings of the ship was routine, and finding the weight of each item was made simple by the fact that each had been pulled off a workboat by the rig's massive cranes, equipped with scales. The hard part for Dave was learning the names of each item. Dave had no direct involvement with most drilling procedures, with the exception of being responsible for storing and then delivering the powdered cement that would be mixed and used in the well. But unlike some of the marine staff, he took an active interest in the process and saw his new role as an opportunity. It allowed him to pull guys aside and ask questions: how equipment was used, how it was handled on deck, and, most important, what it was called.

As he became more familiar with all the complex tools and materials used in deepwater drilling, he was better able to appreciate what the Horizon accomplished in the summer and early fall of 2009.

To the crew it was just another well, one of nearly thirty wells it had drilled since the rig floated. One well was pretty much like the next. There was the derrick-topped, crane-studded, 200-foot by 300-foot rectangle of the rig itself, and the 360-degree panorama of the Gulf of Mexico. One patch of ocean surface looked the same as another. As they moved from drilling one well to drilling the next, without a GPS reading you wouldn't know the rig had gone anywhere.

All that was true of the well they drilled in the Tiber Field, a deepwater oil site located in the Keathley Canyon in the Gulf of Mexico, 250 miles southeast of Houston. The work, the drilling and construction of a well structure beneath thousands of feet of ocean, was like every other well the Horizon had worked. Except that it kept going on, and on. By the time the drilling team had hit its objective, they had reached 35,000 feet beneath the mud line and 39,000 feet below the ocean's surface, the deepest well ever drilled.

When it was over, most of the guys on the rig just felt relief. They wouldn't miss the two-hour helicopter commute out to this more than usually distant spot. Some did feel pride at the congratulatory e-mail messages sent to the crew, and the words of praise from both Transocean and BP managers.

But the accomplishment may have seemed even more significant to those who had no idea what it took to drill a well in ultradeep water. As the news was broadcast around the globe—the deepest well ever!—it was taken by many as particularly stunning evidence that technology had conquered the monumental challenges and hazards of offshore drilling. The Tiber Field triumph was an exclamation point to the rising sentiment that offshore drilling had become so advanced, it was virtually fail-proof.

CHAPTER TEN

LATCHING UP

February 2010

Block 252, Mississippi Canyon, Gulf of Mexico

Nature provides no shortage of elaborate, even bizarre mating rituals. The fierce head-butting of elk bulls in rut, the four hundred distinct mating chirps of the grasshopper, even the female praying mantis's habit of decapitating the male during copulation and ingesting his head—all are astonishing in their own right, but they pale in comparison to what happened above the Macondo well on February 9, 2010.

After the Marianas was towed away for repair on Thanksgiving Day, it took some tinkering and horse trading and a few more lost months before Transocean's drilling schedule could be rearranged and another company asset could be redirected to Block 252 in the Mississippi Canyon. That turned out to be Deepwater Horizon, which was the first available of the twenty-two Transocean ships or rigs capable of drilling in ultra-deep water. The dynamic positioning operator on the Horizon's bridge input the GPS coordinates of the Macondo wellhead, and the computer pointed the rig in the right direction and fired the thrusters. That part was simple. But after the Horizon reached its destination, things got considerably more complicated.

The Marianas had left Macondo a 3,900-foot-deep, steel-lined hole to nowhere. The well was less than a third completed and topped by a metal funnel that stuck up just above the ocean floor. The assemblage resembled a supersize version of one of those funnel and tube combinations used to pour gas from a five-gallon jug into the empty tank of a stranded car. In this case, the purpose of the funnel, otherwise known as the wellhead, was to receive the protruding 27-inch diameter male end of the 325-ton blowout preventer dangling from the end of a 5,000-foot steel string attached to a vessel bobbing in the waves high above.

To conceive how difficult it is to drop the BOP stack's connector pipe into the well's hole, first imagine standing on the observation deck of the Empire State Building and attempting to lower a soda bottle at the end of a 1,200-foot-long string into a garbage can on the sidewalk. It's extremely windy, and you're wearing roller skates. Now consider that, with the building encased in clouds, it's impossible to see the sidewalk, much less the garbage can.

Imagine an observer with a cell phone at the bottom giving directions as the bottle descended. Every motion made by the person on the observation deck would take time to translate down the long string, and the effect on the bottle of his movements interacting with the swirling winds would be virtually unpredictable.

But all of that would be easy compared with what the crew of the Horizon was attempting to accomplish. Instead of a thousand feet of insubstantial air to contend with, they were dealing with 5,000 feet of water exerting 2,300 pounds of pressure on every square inch of surface area. An assistant with a cell phone wearing even the most advanced scuba gear would be dead before he got a fifth of the way to the bottom. Even a nuclear submarine would be crushed like a grape about halfway down. And instead of simply unspooling a string, the Horizon's drilling crew would have to assemble, piece by piece, 75-foot segments of 19½-inch-diameter steel pipe weighing more than 30,000 pounds each, and feed them slowly through the center of a moving rig.

Before any of that happened, they had to locate the wellhead. This wasn't as easy as motoring to the coordinates locked into their GPS system. The GPS coordinates referred to a point on the surface of the ocean and were all but useless in locating a precise point 5,000 feet down. For any practical purpose, there was no "straight down" in the Gulf of Mexico. "Straight" was a theoretical concept rendered all but meaningless by the constantly swirling currents and the prodigious distance, just as there was no dropping a bottle on a string "straight down" from the Empire State Building.

But the Horizon could do a lot better than an assistant with a cell phone.

The rig spent a few days over Macondo taking the opportunity to inspect and maintain its nine-year-old blowout preventer, the same one that had launched with the rig from Korea. Not only did the crew's livelihoods depend on a properly functioning blowout preventer; so too did their lives. But the only time it could be inspected and maintained was when it was on deck between wells.

Once the subsea engineers were satisfied with the condition of the BOP, they got to work on the elaborate preparations necessary before they attempted to lower it to the wellhead. First the marine crew hooked acoustic beacons to the grating of a steel cage about the size of something that might have been used to punish a troublesome prisoner in a Soviet gulag. The cage was then lowered over the side by a large winch spooled with ten thousand feet of umbilical cable. The cable whirred off the spool interminably. It took a full hour before the cage had descended nearly a mile, to one hundred feet above the sea bottom. Finally the winch yanked the cable to a halt. The cage door opened and something very like a giant hermit crab snuck out. This was an ROV, or remotely operated vehicle. The ROV was equipped with powerful head lamps, video camera eyes, and mechanical arms that gave it that crab look. One of the arms had a cutter where a crab's pincher would be. The other arm had a grabber, which could be manipulated to grasp and use a variety of simple tools, like a scooper, a squeegee, or a pressure washer.

An umbilical cable with the ROV's power, hydraulic control, and video lines snaked back up to the rig and connected to the ROV shack—basically a shipping container furnished with comfortable chairs, video screens, telephones and, best of all, joysticks. The joysticks not only controlled the ROV's movements, but also operated the lights, camera, and the machine's arms.

Being an ROV operator was a video gamer's dream job—a 3-D entertainment that wasn't virtual, but real. Not surprisingly, the ROV operators tended to be young video game enthusiasts who could turn the ROVs in intricate loops and work wonders with the mechanical arms. Either that, or they were older men, invariably small in stature, who had gotten wet to drive the submersibles themselves back in the 1990s, when anchored rigs worked comparatively shallow water and the tiny machines were manned. Either way, both the old and the young operators were geniuses at what they did, and the rig would be nearly helpless without them.

For the moment, the ROV's task was to locate the wellhead and surround it with five sonic buoys that would allow the dynamic positioning operator to pinpoint its exact location relative to the surface. When the ROV emerged from its cage, the operator switched on its sonar. Because of the currents buffeting the crane cable on the trip down, there was no telling where the ROV had ended up, or even what direction it was facing. They watched the sonar screen as the ROV slowly rotated. A blip appeared—almost certainly the wellhead. They switched on the lights and the camera, and drove the submersible in the direction of the sonar blip. The powerful lights on the ROV pushed back the total blackness of the depths. After a few minutes that seemed much longer, the bulk of the wellhead appeared on the screen. The ROV closed the remaining distance and scurried around the wellhead, inspecting it from all angles with its camera. Grasping the pressure-washing tool, the ROV sprayed off the mud that had settled atop the steel. The ROV operator peered at his video screen, looking for the carpenter's bubbles welded to the wellhead. If the wellhead had sunk unevenly into the mud, the BOP wouldn't be able to latch on properly, and the whole effort would have been wasted.

When the bubble appeared on the camera, it was squarely within the level lines. All was as it should be. The ROV could swim back to its cage.

Now the Horizon slowly motored a thousand feet north of the wellhead and hovered. The ROV reemerged and focused its cameras on the cage and the acoustic beacons hanging from the side. They were four-foot-tall steel tubes, four inches in diameter, each containing an extremely sensitive underwater microphone, and each spliced to a rope anchored with a cement block. The ROV used one hand to grab the spliced rope and the other to cut the line securing the buoy cylinder to the cage. Then, still grasping the rope, it swam down to the bottom, placed the weight on the seafloor, and let go of the rope. It returned to its cage, the rig moved to another position a thousand feet from the wellhead, and the procedure was repeated until the wellhead was surrounded by a pentagon of sonic buoys.

Only three of the buoys were required for the triangulation that would give an accurate position of the wellhead. The fourth was there in case one of the other sonic buoys stopped functioning, and the fifth was there to back up the backup. Nobody wanted to have to repeat this process.

When all the buoys were in place, the ROV was cranked back to the surface, and the rig's transponder, hanging from a pole beneath the pontoon, began to ping the five buoys in sequence. With each ping, the lock on their positions grew slightly more precise. The calibration was complex and dependent on water temperature, salinity, and a host of other variables. For it to succeed, the sea had to be completely quiet. If a workboat was tied to the rig delivering supplies, it would be sent out of range; if workers were striking hammers they would be asked to stop. Sometimes a whale would be cruising the region, making an underwater racket. The whale couldn't be ordered away, of course, so the crew just had to wait for it to leave on its own. Nothing could be allowed to drown out the distant echo of the beacons.

The calibration was painstaking and complex, but the hard part wouldn't really begin until it was complete.

The first step was to get the BOP in position, suspended above the water over the rig's "moon pool"—the hole in the center of the rig directly beneath the derrick that might, hypothetically, reflect the moon on a clear night. The crew unbolted the BOP from the deck, where it had been fastened since being hauled up from the last well, sitting on top of a hydraulically powered cart that looked like a half-scale flatcar. Now the cart inched along a track carrying its enormous burden toward the edge of the moon pool, where the

Deepwater Horizon, one of the most powerful industrial machines ever built.

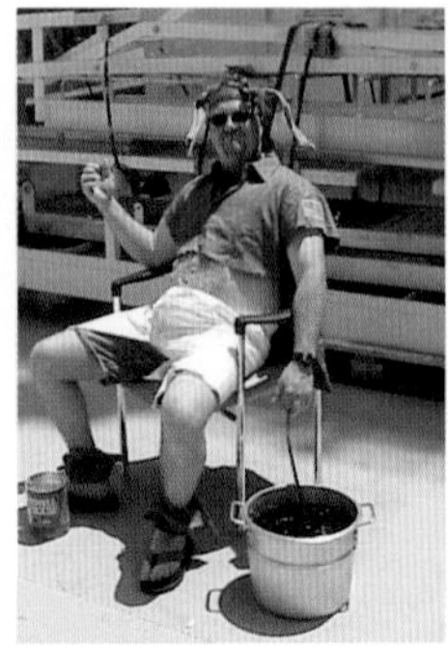

Above, from left: Doug Brown at home with his wife, Meccah, and daughter, Kirah. *(Courtesy of Doug Brown)* Jason Anderson as the "Sea Baby" during the Horizon's "Shellback" ceremony in 2001. *(Tyson Cullum)* Dave Young in his dorm at SUNY-Maritime, circa 1995. He expected to sail a ship, not an oil rig. *(Cindy Konrad)* Captain Curt Kuchta on the Horizon, circa 2009. *(Courtesy of Curt Kuchta)*

Crane operator Dale Burkeen, who would become one of eleven casualties, with his son, Timothy, and daughter, Aryan. *(Courtesy of Janet Woodson)*

Clockwise from left: Daun Winslow, one of the VIP visitors on April 20. *(U.S. Coast Guard)* John Guide, BP's well team leader for the Horizon. *(U.S. Coast Guard)* Randy Ezell, senior toolpusher. *(U.S. Coast Guard)* Jesse Gagliano, technical advisor for cement work on the Macondo well. *(AP Photos)* Mike Williams, chief electronics technician. *(U.S. Coast Guard)* Stephen Bertone, chief engineer. *(AP Photos)*

The original Deepwater Horizon supervisory crew in 2001, as the rig floated in the harbor at Ulsan, Korea. Doug Brown is the fourth man from the left in the second row. Two of the victims of the blowout nine years later are also pictured: Donald Clark, first on the right, front row, and Jason Campbell, fifth from the right, second row. *(Doug Brown)*

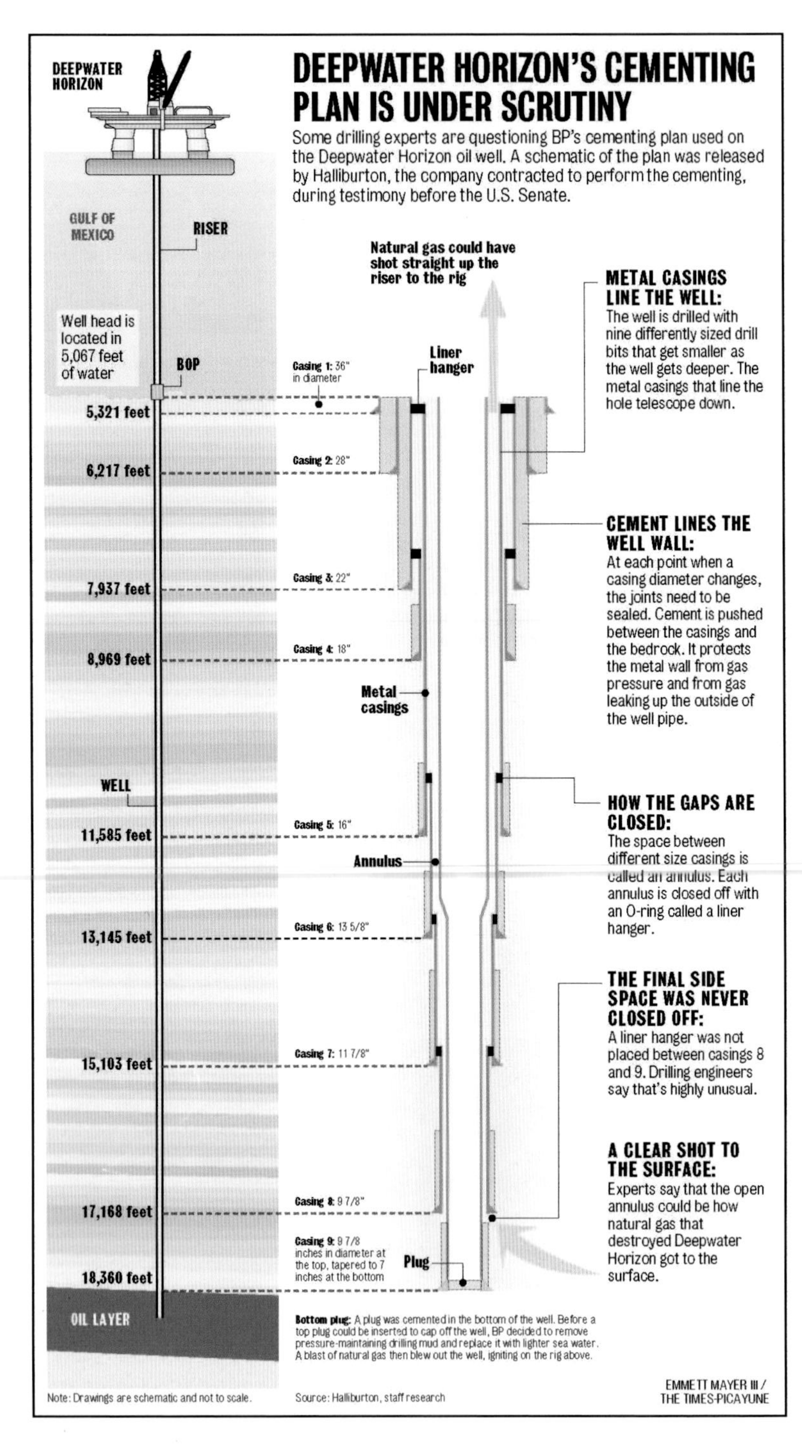

(Times-Picayune)

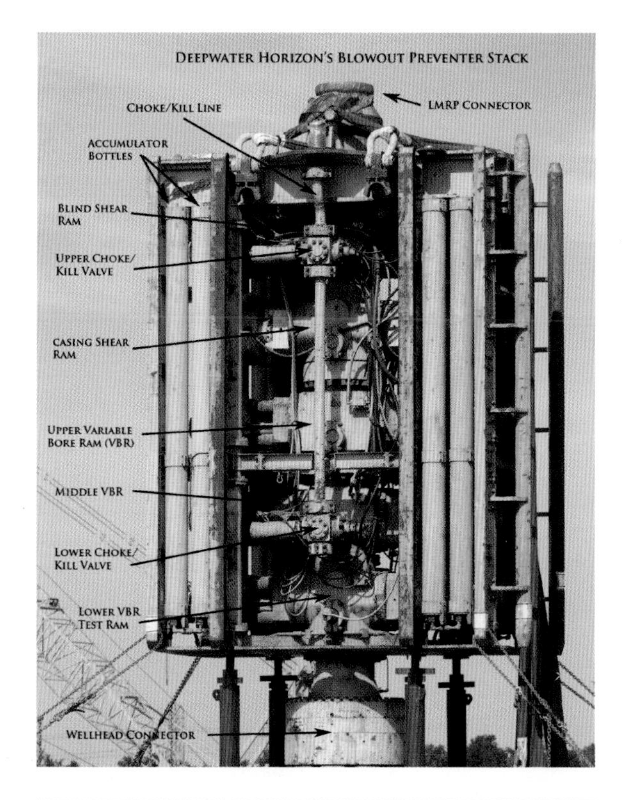

The blowout preventer. *(Robert Almeida)*

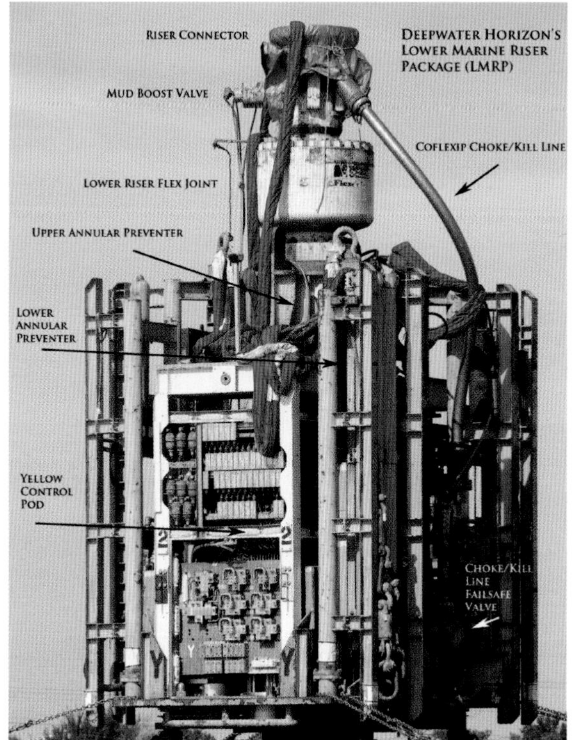

The lower marine riser package. *(Robert Almeida)*

With inexhaustible fuel from the blown-out well, the fires on the Horizon raged for thirty-six hours. *(Getty Images)*

The Horizon's helipad is visible in the foreground, directly above the rig's bridge, and just to the right of the empty lifeboat berths. *(Getty Images)*
Below: On the morning of April 22, the combined effects of fire and firefighting overcame the Horizon's ability to float. It listed, then capsized and plunged to the bottom. *(AP Photos)*

(AP Photos)

tracks took a left turn out over thin air. A gantry rose up to encase the BOP and lock on from either side.

The gantry lifted the BOP off the hydraulic cart, then out toward the center of the moon pool, directly beneath the derrick, and held it there. It was the mirror image of a rocket preparing to launch. Instead of emerging from the gantry and soaring toward the moon, it would be descending into the moon pool, then launching into the inner space of the ocean.

While the BOP hung above the water, its connector pipe aiming toward the wellhead, the drilling crew had been readying the first section of riser pipe.

With its dynamic positioning system, the rig was astonishingly capable of keeping itself on location, but the stability couldn't be perfect, given the ocean's heaving waves and currents. The steel riser dangling five thousand feet below the rig would need flexibility to manage these forces. So at the top of the BOP, below the first section of the riser, the crew would install a coupling called a flex joint, a substantial steel and rubber coupling that allowed the rig to move off station by a certain amount without damaging the riser or BOP. The flex joint, which alone weighed several tons, was hooked to the top drive—the huge engine block that, during normal drilling operations, slid down the derrick, turning the drill as it drove deeper into the earth.

The top drive was operated by the driller from a small office directly beneath the derrick, called the drill shack. It used to be that the drill shack was a gritty place where the driller manipulated levers and foot pedals to control the top drive from a ratty chair, but in a fifth-generation rig like the Horizon, the driller sat in air-conditioned comfort at an ergonomically designed console, overlooking the drill floor behind a wall of glass. He operated the top drive through the rig's computer system using a joystick. On his computer screen he could see virtual dials and gauges that were hooked into the rig's extensive sensory network and told him whatever he needed to know about what was happening on deck or in the well. His need to know was critical. The sensors could alert him to a well that was about to blow out, or one that was crumbling because of too much drilling pressure. It could also give him early warning of a malfunction in the dynamic positioning system.

Besides failure to anticipate a major storm, the two biggest threats to the system keeping the Horizon steadily above the well were either a sudden power blackout or a power surge. In a blackout, the engines would shut off and set the rig adrift. In rarer instances, a computer glitch or mechanical breakdown could ignite a "drive-off"—powering up one or more of the thrusters from 20 percent to 100 percent in an instant, pushing the rig off the well and threatening to rupture the riser.

If for any reason, the rig begins to move out of position, the system signals a warning. At a certain distance, a yellow light goes on—this is 50 to 60 feet from center in 5,000 feet of water. When the light flashes, the subsea engineer will stand by for an emergency disconnect, known as EDS. Drilling will stop and the drill pipe will be pulled up a little to make sure shear rams in the BOP have nice, smooth pipe to cut into rather than the thick joint between pipe sections. Then the driller will watch his panels. At 100 to 120 feet from center, the red light comes on. That's when there's no choice. The EDS button has to be pushed to prevent rupturing the riser.

Nobody wants to push the EDS button. A disconnect is expensive—it would take a day or two before the rig could reconnect to the well and get back to work. But failure to disconnect can be even more costly. The BOP could be damaged or destroyed; the whole wellhead could tip over. It could cause a blowout.

Some rigs have a disconnect a year, and others never have one. For the driller, emergency situations are always in the back of the mind, but almost never in the forefront. Day to day, he is mostly concerned with the operation of his most powerful tool, the top drive.

For now, all it had to do was pull the flex joint up the derrick, then lower it atop the suspended blowout preventer. Drill hands in a hydraulic lift rose to the junction of BOP and flex joint, then bolted them securely together with impact wrenches. Another short piece of riser pipe, which would soon be essential, was bolted to the top of the flex joint. The top drive rose a few inches, taking the weight from the gantry, which disengaged and slid back along its track, out of the way. Now the top drive lowered again, until the short riser joint was sticking a few feet above the top of the drill floor.

It was time to install the spider—a heavy scaffolding with legs that spanned the drill floor above the moon pool. The center of the spider fit around the protruding short section of riser pipe, and powerful clamps held it in place. The driller lowered the top drive a few inches, shifting the weight of its load onto the spider, which caught and held. The top drive could be disengaged and raised back up the derrick, leaving half a thousand tons of heavy-duty plumbing hanging seventy-five feet above the Gulf of Mexico from something resembling a daddy longlegs made of steel.

To get this far had taken about six hours. Now they could run the riser.

The Deepwater Horizon had enough riser pipe on deck to run down into seven thousand feet of water. It was stored on the riser deck, a space the size of half a football field immediately aft of the end-zone-sized drill floor. The pipe was stacked twenty feet high in rows against perpendicular stanchions and beneath the riser crane, which hung off an I-beam running port to starboard across the riser deck.

Fresh from the huddle of the pre-job safety meeting, the team deployed. The crane operator climbed up into the crane cab; the deck foreman found a spot where he could see the entire deck at once; the roustabouts strapped themselves into their harnesses, climbed the rungs of the riser deck stanchions, then jumped off onto the stacked pipe and tied off. There, balancing on the top riser pipe, they reached for the crane blocks—there were two, one on each side—and connected one to each end of the first riser section. The crane lifted the pipe to float horizontally above the stacks, then rolled itself to the middle of ship, swaying the pipe until it lined up with the drill floor. Roustabouts grabbed what would become the bottom end of the pipe, seated it on a skate, and hooked it to a winch cable. The winch pulled the pipe on its skate along the drill deck until it was directly beneath the top drive. Now the far end of the pipe was attached to the top drive and was lifted off the deck, rising at an increasing angle until it dangled vertically above the spider.

From here it would be connected to the pipe end clamped in the spider, the spider would release, and the top drive would lower the now dangling section seventy-five feet. Then the end of the new section would be clamped in the spider, the top drive would release, and the process would be repeated seventy times.

But first, they had to wait.

Up to this point the BOP was still dry. Lowering the next section of pipe would put it in the water. The force of the current pulling on the BOP's enormous mass could push it against the side of the rig or break the spider's clamps or otherwise create havoc.

To prevent that from happening, the rig would have to be moving exactly with the current as the pipe was lowered, section by section. Careful calculations had been made by the bridge crew. It would take twenty-four hours to run the riser to full depth. If current was running at about a half knot, and the rig motored twelve miles up current, they could drop the BOP in the water and drift with the current as the riser descended. The rig, the BOP, the riser, and the current would all move in unison, and by the time it got to a depth of 5,000 feet, it would be right over the wellhead.

In anticipation, the rig had been under way throughout the preparations. When the rig arrived at the calculated point, one of the marine crew came up on deck and used some low-tech seamanship. He cast his eye around the surrounding sea, looking for a piece of driftwood, a clump of seaweed, anything that drifted in the current. Then he watched it for a few minutes. If the clump and the rig stayed in synch, he gave the driller the thumbs-up and the driller gave the command.

"Let's get her wet."

The top drive rode down the derrick until the BOP splashed in the water. Then the driller slammed on the brakes. A crowd stood around the moon pool and watched the protruding pipe for five minutes. It just hung there, straight as an arrow. The driller flicked his joystick and the top drive dropped another ten feet, then stopped again. The riser was still straight down the middle of the moon pool. They were in the clear.

The drift strategy did have one built-in danger. If the captain and chief mate hadn't plotted carefully, or the crew was working faster than anticipated, the dangling riser could bang into a submerged ridge or subsea mountaintop. The collision, to say the least, would not be healthy for the BOP.

This had been known to happen. A couple of years earlier on the Discoverer Spirit, the drill floor had been making great time, which meant the BOP was hanging farther down than it should have been when the rig passed over a ridge. The ridge was mostly mud, so no real damage was done as the BOP plowed through it, but the stack had be pulled to the surface and inspected. The Transocean captain was so mortified, he submitted a letter of resignation. The company refused the resignation but gave the captain a warning, and instituted a new policy to ensure that no one ever topped his record for the deepest grounding of a ship . . . ever. So, from then on, whenever a riser was drifting in the current, an ROV would be deployed to hover one hundred feet below the submerged BOP, scouting for obstacles.

Most often the plan was sound, and there was nothing to watch for. The ROV operators would get bored and start looking for sharks. If they spotted one, they'd send out an alert on the PA system, "Turn on your TVs, we got a big, toothy one smiling on the ROV channel." It was like a scene out of the movie *M*A*S*H*, and a whole lot more mature than some of the other things that were occasionally sent out over the PA, like turkey calls or amplified farts.

If no sharks showed up, the ROV crew could find other diversions, like the magically shrinking Styrofoam trick: Write a wife or girlfriend's name on a Styrofoam coffee cup, attach the cup to the ROV cage, send it into the deep. When the ROV was brought back up to the surface, the cup, subjected to extreme pressures from all angles, would return in perfect shape, but the size of a shot glass.

Or maybe someone would have been bragging about the new

watch he'd just bought. You could always find out if the ads claiming that some titanium timepiece would keep ticking at any depth are for real. (They aren't.)

As the ROV swam below, the thrusters were firing only to keep the rig's bow into the current. If something went wrong with the riser deployment—a wrench broke, the spider clamp jammed—the dynamic positioning would have to be reprogrammed to hold position until the problem was fixed, leaving the riser temporarily vulnerable to the current. To compensate, the rig's heading and ballast system would be adjusted, intentionally causing the rig to lean to one side, giving the riser room to be pulled at an angle with the current.

Stopping every ten joints to pressure-test the assembly and make sure there were no leaks, it took two twelve-hour shifts before the Horizon arrived above the wellhead. But now that the climax approached, almost everyone near a TV was keeping an eye on the ROV channel for the dramatic conclusion.

The BOP was hung off a couple hundred feet above the wellhead. The calculators were pulled back out to make sure the riser length was as close to perfect as possible. If it was ten feet off, they could install special short lengths of riser called pup joints to make up the difference.

Then it was time to connect "the jewelry"—beginning with a second flex joint at the top of the riser to match the one at the bottom. The two flex joints compensate for the slight horizontal movements of a working rig. But the rig is also subject to the heave of the swelling ocean. For that, they installed the telescoping joint, called a slip joint, a $3 million tube within a tube that can expand or contract up to fifty feet with each passing wave. About a half-dozen thick steel cables called tensioners were then connected to the outer pipe of the slip joint and run up in every direction, to the top of the moon pool and out to the side of the rig, where they were wrapped around two wheels and fed into hydraulic winches set to pull the cable in or play it out, maintaining a constant tension on the top of the riser. If the rig moved up thirty feet, it would play out thirty feet of cable, and when it fell back down, the cable would be retracted. Like a potbellied man's suspenders, the tensioners kept the riser string from slacking off below the rig.

Now that the jewelry was connected, all five thousand feet of riser and the BOP dangled above the wellhead. All that remained was to drop the bottle in the garbage can.

The ROV operator, the driller and the DPO all were talking to each other on headsets. The rig was oscillating within a six-foot radius above the funnel-shaped wellhead. That motion was being transferred down the riser five thousand feet to the BOP, but there was a delay of as long as a couple of minutes before the shifting force at the surface resulted in a shift in the position of the BOP. The riser string was heaving up and down by as much as ten feet in the waves. As the BOP oscillated above the wellhead, the driller had to pick the exact moment that a downward thrust of the top drive would "land out" the BOP, seating its wellhead connector into the latching mechanism, linking the well, the riser, and the rig as one.

The idea was to wait for the rig to be at top of its rise, then drop the assembly very quickly. As the connector dropped into the wellhead, the driller could lock it in by rotating the top drive, then slack off the weight. But there was only one way to get it right, and many ways to mess up. If the driller dropped the BOP too quickly, it would smash into the wellhead. If he did it too slowly, the rig would rise up again before the BOP could connect. On the plunge that followed, it would jam down on the edge of the funnel. Either way, the collision could damage the BOP or knock the wellhead over. As mistakes go, both were doozies, multimillion-dollar screwups.

And if the driller needed any more performance pressure, a sizable percentage of the crew was watching on the ROV channel.

A latch-up is considered a clean success if a driller locks it in within ten to twenty cycles of heave. Anything more than that is considered a bit of an embarrassment, even though it's also probably the most common outcome. Maybe one out of a hundred attempts results in some significant damage. A driller who's thinking about the one-in-a-hundred chance is going to have a hard time keeping his hand steady.

The Horizon heaved in the waves. The driller waited, waited, then made his move. A mile down, a connection was made. The Deepwater Horizon was off to an auspicious start. Everyone aboard had good reason to think the fog of bad luck that had shrouded Macondo had lifted at last.

CHAPTER ELEVEN

KICKS

Mid-February 2010

Macondo Prospect

Marianas had left the crew of the Horizon a hole eighteen inches wide and encased by steel pipe and cement, extending 3,902 feet below the sea bottom. But they'd also left a hole of a different kind, a financial one. Even before the BOP breakdown and the hurricane, the Marianas had missed every deadline set for it, and after completing less than a fourth of the well, it had used up half the hoped-for budget.

Now it was the Horizon's turn to try to subdue Macondo.

According to the well plan, the rig's first task would be to drill 2,500 feet farther and two inches narrower below the existing bottom and line the hole with sixteen-inch-diameter casing.

From its inception about two thousand years ago in China, the process by which human beings have drilled wells has always entailed impressive engineering—often unprecedented. In an astonishing spurt of ingenuity, driven by a desperate desire to retrieve deep deposits of brine water to produce salt—as essential to the ancient Chinese as oil is to us today—salt miners developed many of the basic strategies used by modern drill hands. They built a derrick of bamboo to support a pulley system that allowed them to raise and drop a bamboo pipe fit with an array of heavy iron drill bits weighing as much as five hundred pounds or more. Several men would stand on a wooden plank set on a fulcrum to raise the drill string, then jump off to let it fall, pulverizing the rock in the well hole. When enough debris accumulated in the well, the drillers would drive down a hollow pipe with a leather flap on the bottom, which acted as a valve. The debris would push up into the pipe, then as the pipe was lifted from the hole, the weight of the debris would push the flap closed, keeping the dirt from falling out the bottom. As the well got deeper, more bamboo segments would be attached to the drill string. If the well walls began to crumble, segments of hollow tree trunks were driven into the hole to act as a casing.

The method used on the Horizon retained an impressive similarity. Bamboo has been switched out for steel in the derrick and the drill pipe, and the drill bits are diamond-encrusted, multiheaded swivels with holes spaced regularly along their extensions that spurt oil-based drilling lubricant called "mud." Instead of just letting the drill bit fall, the 112,000-pound top drive assembly can spin and drive the drill bit into solid rock at anywhere from 40 to 200 revolutions per minute while exerting a downward pressure of up to 20,000 pounds or more. There are a wide variety of bits for different uses and conditions, but one deep-sea bit, encrusted with synthetic industrial diamonds (diamonds created in a laboratory but with all the hardness of natural diamonds), can cost over a million dollars.

As the diamonds in the bit bore into the rock, the top drive descended down the derrick, pushing and twisting the drill pipe farther into the earth. The liquid drilling "mud" forced through the channels of the drill bit pushed the debris back up the well bore to the surface, just as a raging river carries debris downstream. When the top drive had lowered all the way to the rig floor, it was disconnected from the pipe and raised back up the derrick to await a new 93-foot section of drill pipe, made of three 31-foot individual pipes screwed together, end to end. These assembled sections hung vertically on racks in dozens of rows ten pipes deep, waiting for the driller to thrust them into the hole. Like a massive wind chime, the pipes sang as wind and waves rocked the rig. When the driller was ready, he manipulated his joystick to swing a pipe-handling machine over to the racked pipe, grasp one stand of pipe in its grip, and swing it over the open hole, directly beneath the top drive. Then he let the top drive fall gracefully on its block until it connected to the top of the stand of pipe. A floor hand scooted in with a bucket of the thick grease known as pipe dope, which acted as both lubricant and sealant. He slathered the dope on the pipe's threads with a paintbrush, then signaled the driller, who lowered the top drive until the bottom of the new pipe met the top of the stand already in the hole. A twist of the driller's hand rotated the top drive, which turned the pipe until it was firmly connected to the string. Then the drilling resumed.

Depending on the situation, drill pipe weighs anywhere from about 1,500 pounds per 93-foot string, to more than three tons, so by the time the bit reached the Macondo oil deposit, the top drive would be rotating and driving from 200 to 800 tons of steel pipe.

Everything about drilling got more difficult and took longer the deeper the hole went. The weight of the drill string grew incrementally, as did the force needed to turn it. The heat and pressure in the hole increased with every foot of depth. Each time a drill bit had to be changed, the thousands of feet of drill string had to be hauled up and disassembled section by section, then reassembled and lowered back into the hole. As a result, to conserve energy and resources as the depth increased, these deep wells were built like inverted wedding cakes, with the widest sections at the top and increasingly narrower sections below. The narrower the section, the less time, energy, and materials it took to drill, case, and seal it.

The hole itself was just the first step of well construction, the equivalent of a pit dug in the ground for a foundation. When they reached their intended depth, 2,500 feet beyond where the Marianas had stopped drilling, they pulled the drill string up and prepared to lower a 16-inch hollow steel pipe into the hole. Even though the new part of the hole was 2,500 feet long, the 16-inch casing would be much longer, running inside the previous section of the well, from near the top of the 22-inch-diameter pipe that had been installed months earlier by the Marianas to the bottom of the new section. A protruding lip at the top of the 16-inch casing would land on a ledge built into the 22-inch casing, and hang there with only the force of gravity holding it in place until it was cemented and sealed. The drilling schedule called for this section to be completed in six to seven days. But from the beginning, the work fell behind.

Drilling wells is rarely a smooth procedure and after the drill team crossed the 4,000-foot threshold, Macondo started to "kick." Even small amounts of gas, deep in the earth, expand exponentially as they rise to the surface. A few cubic feet at depth could become enough gas to fill the Superdome at the surface. The explosive expansion of the gas creates tremendous force and pushes everything out of the hole before it. As the Horizon's drill team bore down—still nowhere near the oil and gas deposits highlighted on the geological surveys—they nonetheless encountered small pockets of gas flowing into the well unexpectedly.

In small doses, with enough warning, kicks can be controlled.

Jason Anderson had become an expert in doing just that. Since Jason had come on the rig in Korea, he had worked his way up through the ranks—from a pump man in the mud storage pits to the assistant driller, to driller, to toolpusher—just as everyone had thought he would. At thirty-five, he was essentially a drilling foreman and was headed higher still. Life on the rig wasn't just a job to Jason; it was a calling. He'd made a point of studying the ways a well could act up, and the art and science of keeping it under control. He knew enough that Transocean had offered him a job as an instructor at their well-control school. Jason was tempted. He wanted to keep moving up. But a teaching gig, a "land job" based in Houston, meant leaving the rig behind and Jason wasn't ready to do that. He had his sights on senior toolpusher, then OIM. He wanted to do it all.

For the three weeks Jason spent at home every cycle, he liked to kick back, hang out with his family, play golf (badly) with his friends, go hunting (five-year-old Lacy was already talking about coming with Daddy to "boom" her first deer), or maybe take off in the camper his wife, Shelley, always kept stocked and ready to go. He, Shelley, Lacy and little Ryver, who was just starting to walk, could pile in and wander the Texas outback until it was time to fly back to New Orleans to catch the crew copter for another hitch. The brick ranch house on the rural lane in Midfield, Texas, near the Gulf Coast, was his real home, but the rig on the Gulf of Mexico was something like home, too. He spent as much time with his crew as he spent with his family, and he felt almost as essential to their well-being. Especially on jobs like Macondo.

Kicks were always troublesome, but Jason knew what to watch for and how to respond. In some ways, he was like a detective.

The drilling team worked blind, essentially jamming sticks down a deep, dark hole. They could only know what was going on indirectly, by making deductions from the evidence of their measurements—and the rig and its drilling equipment were designed to measure just about everything. One of the biggest clues available was always the drilling fluid, called "mud," for the way it looks and feels. Mud is the lifeblood of the drilling process, and is watched over and obsessed about by "mud engineers." Nobody on the rig ever so much as smiled at the job title. On a rig, mud was serious business. An oil or synthetic oil-based concoction, it contained barite, or barium sulfate for weight, as well as various chemical additives to tailor it to specific uses and make it environmentally friendly. The mud filled the hole, preventing the walls from caving in. It also served to cool and lubricate the bit as it was forced out of jets at high pressure to carry the drilling debris back up and out of the well. The contaminated mud is processed so it can be reused, but also to learn from the debris. One of the most critical things the debris can indicate is the presence of gas, which is a flashing caution light that a well has ventured into a hydrocarbon zone.

But the first sign of a kick is often seen in the *amount* of mud that returns. The well bore is filled to the top of the riser with mud. Like a full glass of water held under a faucet, any amount of new mud pumped in should be matched by overflow coming back out. The mud fluid is measured as it goes into the well, pump stroke by pump stroke, and measured again when it comes flooding back out into storage areas called mud pits, which occupy a large portion of the lower deck, just aft of the derrick.

If more mud comes out than was pumped in, that could only mean that something down in the well is pushing back, forcing the mud up and out. That would be the kick. Kicks are fairly common and occur on almost every well. They are nuisances that can become disasters if they aren't monitored closely and managed adroitly.

Jason was intent on doing both. At the first sign of a kick, he could activate a feature of the blowout preventer called an annular preventer, a (very large) steel-reinforced rubber doughnut that squeezed tightly around the drill pipe and sealed off the space around it. This allowed the well to settle, a strategy similar to capping a soda bottle that's about to fizz over. It also gave the crew time to pump heavier mud into the well. Mud is significantly heavier than water—which weighs about 8.3 pounds per gallon—and can be made even heavier depending on the additives put into it. Ultimately, it can weigh nearly twice as much as seawater. When you multiply by the thousands of gallons it takes to fill an 18,000-foot hole, that's a very considerable weight, and usually enough to counter the upward force of oil and gas trying to push to the surface. It's a very straightforward equation: The downward pressure of the mud has to equal or exceed the upward pressure of the hydrocarbons seeking to escape.

Proper use of these tools controlled the kicks. But all these maneuvers were complicated and took precious time.

Around the middle of February, when it became clear the sixteen-inch pipe section was seriously behind schedule, yet another setback for this Macondo project, the BP company men got itchy. "Let's bump it up," one of them said. Jason interpreted that as an instruction to push the drill harder and faster, which could get them through this troublesome section more quickly and hold down the escalating cost. But going faster meant exerting more pressure against the geological formation. And sometimes more is just too much. The terrain traversed by a well is as varied as terrain aboveground. It can range from dense, impermeable rock to pressure-compressed sand that can easily crumble when pushed too hard—which is exactly what happened.

A few days after the company man exhorted the drill team to bump it up, the bottom very literally fell out. The first sign of trouble came once again in the mud in/mud out calculus. Only this time, instead of too much mud returning to the pits, there was too little. Somewhere the walls of the well had given way and the mud was escaping into the surrounding geology. This was not good. The collapsed wall was a structural weakness in the well. But it also meant that barrels of mud were washing away. Despite its name, mud wasn't cheap. In fact, it cost far more than refined gasoline, between $200 and $500 per 42-gallon barrel. Formation collapse was called a "lost circulation event" on the rig, because the loss of circulating mud was how it was diagnosed. Thousands of barrels' worth of mud could escape when a well wall failed, so the mud loss alone could very quickly became a million-dollar problem.

Lost circulation could be controlled by pumping even more mud into the hole, this time containing thick and/or sticky additives—including items as humble as ground-up peanut and walnut shells. This "lost circulation material" is plastered against the walls by pressure, forming into a kind of patching material over the gaps.

But pumping and plugging took time. Between the cost of the lost mud, the cost of the replacement material, and the time it took to diagnose and patch the leak, just this first section of the well was on its way to being two weeks late and at least $14 million over budget.

Macondo was beginning to pick up the sobriquet that drillers commonly bestow on the particularly incident-prone holes they drill—"well from hell." They don't usually mean too much by it—just another shorthand for the generic gripe of men doing a hard job against stubborn difficulties. But some in the Horizon crew began to take the term seriously.

In late February, an ROV was cruising around the wellhead when an operator noticed something on his video monitor. There was a definite spurt coming from a joint in the hydraulic lines leading into one of the two BOP control pods. These pods, like boxes perched atop the BOP, were the modules that linked back up to control panels on the rig, and through which the BOP could be directed, whether it was opening or closing a fluid line or activating one of its hydraulically powered rams to seal the well in an emergency.

The hydraulic leak was reported to the senior BP company man, Ronald Sepulvado. Of the four company men assigned to the Deepwater Horizon, two at a time, Sepulvado was probably the most experienced, having been with ARCO oil and gas company for twenty years before it was purchased by BP. He'd worked for BP for twelve years, the last seven and a half aboard the Deepwater Horizon. He knew the rig, he knew the crew, and they trusted him to do the right thing, especially when their safety was concerned.

Sepulvado discussed the hydraulic leak problem in a morning conference call with his BP supervisor, John Guide, a fifty-two-year-old engineer at BP's campus on the western outskirts of Houston. Guide had been the Horizon's well team leader for the twenty-four wells leading up to Macondo and he knew the business. Any issue involving the BOP definitely got Guide's attention, and everyone else's. Federal regulations regarding them were strict and explicit. They stated that any rig encountering "a BOP control station or pod that does not function properly" must "suspend further drilling operations until that station or pod is operable."

Fixing the leak almost certainly would have required not only stopping drilling, but pulling the BOP up on deck—another delay of weeks, possibly months. After some discussion, Guide concluded that the leak concerned the least critical element of the BOP, the test ram—which was used to close the system off so pressure tests could conveniently be performed on various parts of the well. As for the continuing loss of hydraulic fluid, the leaky valve need only be turned to the neutral, or "block" position, and it would stop. Subsequent tests seemed to indicate that the rest of the BOP still functioned.

Guide decided that the leaky valve did not meet the standard—not functioning properly—as stated in the federal regulation. Therefore, he concluded, he didn't need to report the leak to federal regulators, and the Horizon didn't need to suspend drilling. The subsea crew set the valve on the leaky joint to "neutral," or "block." This meant that it would be depressurized, and without pressure, the leak would stop spewing fluid. If they needed to use the test ram, it would have to pressure up, which took some time. But the loss of fluid during a limited use would be negligible.

On a huge rig with so many complicated moving parts, these kinds of decisions were constantly being faced. A piece of essential equipment would start wheezing, in one way or another, but it could still be used. To get at the root of the wheeze would be time-consuming and expensive and often could grind operations to a halt. So did you just work around the issue until you could pause for repair? Or shut everything down?

In simplified, everyday terms, it was a little like driving a car that is burning oil. You can put it in the shop and pay to rebuild the engine, or wait until you have more time and money, and in the meantime just keep driving, dump in a quart of oil every time you fill the gas tank, and cross your fingers.

Mike Williams was an ex-marine who had become the Horizon's chief electronics technician around the time they drilled the deepest well, six months earlier. He'd come to the Horizon six months before that, in the spring of 2009. Almost from the moment he'd arrived, he'd been seeing things that alarmed him.

One of his duties as an electronics tech was maintaining the fire and gas detection and alarm system, an extensive network of sensors throughout the rig tied into the rig's mainframe computer. When he arrived, Mike found it in horrible disarray, with many of the sensors not functioning or locked out. As he set about trying to put things to rights, he stumbled on a page deep in the computer for the rig's general alarm. He saw that the alarm had been switched to the inhibited mode, which meant it wouldn't automatically sound if the sensors detected a potentially life-threatening situation. When he reported it, thinking he'd uncovered a serious mistake, he was told that everyone, from the OIM down, wanted it that way, so the crew wasn't awakened at 3 a.m. for a false alarm. They wanted the watch officers on the bridge, who could see fire/gas sensor alerts on their computers, to decide if it merited sounding the rig-wide alarm.

Williams understood the thinking—sometimes a cloud of cement dust could trigger the sensors harmlessly, waking up the sleeping crew members, leaving them drowsy for the next day's long tour—but he didn't agree with the conclusion. Seconds matter in emergencies.

For the last few months, Williams had been struggling with an aging computer system in the drill shack. The system was the driller's window on all the conditions in the well and on the rig, and his control over everything from the mud pumps to the top drive. Out of nowhere, the computer would just lock up—the screen went blue, the "blue screen of death," as it's called. It happened all hours of the day or night. It was more than an inconvenience. When the screen froze, the driller was blind. Williams was told that on an earlier well, the screen went to blue and for a few minutes they had no way to monitor what was happening in the well. By the time they'd fired up the backup computer, they discovered they'd taken a kick.

The system was antiquated, so no matter how heroic their efforts to tinker with it, the threat of a crash would remain. They'd ordered an entirely new system—new computers, new servers, new everything—except software. They couldn't get their old software to run correctly on the new operating system. So they were letting their sister rig, the Nautilus, work out the bugs for them before they installed the new equipment.

The Nautilus was built just before the Horizon, a nearly identical twin except that it did not have dynamic positioning capability. But the drilling mechanisms were almost carbon copies, so what worked on the Nautilus computer system should work on the Horizon's. Meanwhile, they were limping along with what they had.

Sometime in March, Williams had been called yet again from his office near the engine room to the driller's cabin to nurse the computer system. A contractor walked into the back. Cradled in his hands, as if he were carrying a dead bird, was a double handful of stripped rubber. Williams instantly identified it as rubber from the annular preventer—after all, it was pretty much the only rubber down in the well.

Williams glanced nervously at the rubber and said, "What the hell is that?"

"Oh, no big deal. That's normal," he says he was told. "It's not a problem. This happens all the time."

One of the advantages of using the annular preventer was that you could still do some drilling operations while it was closed by gently sliding the drill pipe through the clenched rubber. When used that way, some stripping of rubber did occur, and the driller was careful not to have the annular closed too tightly, or pull the pipe too hard.

But these seemed like awfully big chunks to Williams. Though he would be the first to admit he was no drilling expert, the incident stayed with him.

Then he remembered something: Late one night, not long before he'd seen the chunks of rubber, he'd received a call summoning him to the drill shack. When he arrived, he was told that they had been doing some pressure testing and the annular was closed, and closed tight. Williams saw 10,000 pounds per square inch on the screen. He was asked to investigate whether there had been an input to the control stick that had hoisted the block while the annular was closed.

When Williams asked why they needed to know, he was told, "Well, the block moved about fifteen or twenty feet. We need to know why. We need to know if it was inadvertent stick movement or if it went up by itself."

They eventually discovered it had indeed been an inadvertent stick movement, and Williams now wondered if that mistake had resulted in the extensive hunks of stripped rubber.

Williams didn't know how rubber loss would affect the function of the annular, but he did know there was nothing they could do about it until Macondo was completed and they'd pulled the BOP stack back up on deck.

Within days, Williams was called to the cabin again and told to hurry down. This time it was the BOP control panel. It had gone dead.

Because the driller is likely to be the first to notice the signs of a well about to kick, he needs to be able to activate the BOP instantly, which is why there is a panel in the drill shack, as well as on the bridge and in the subsea supervisor's office. But the drill shack is also directly over the moon pool, and the first place likely to be engulfed in a cloud of gas if a kick gets out of control. So both the drill shack itself and the BOP panel are set up to operate in positive pressure—which means air flow is always out, and never in. That way, even if gas surrounded the shack, it can't enter inside it. And just in case the drill shack was breached and the gas did enter, it wouldn't enter inside the BOP panel, which has its own positive pressure within its glass case. This could be an important consideration because even a small electronic spark can ignite a massive fireball in the presence of natural gas.

What had happened now was someone had held the door to the drill shack open too long, causing it to lose its pressure purge. It was not uncommon for that to happen with all the traffic in there. In just a few seconds, pressure would build back and the purge would be reinstated.

But in this case, the purge system on the BOP panel was faulty, so while the door was open and the pressure seal was lost, the BOP panel detected the lost purge and automatically shut itself down.

By the time Williams arrived, he found the panel was back up, switched by an assistant driller to bypass mode. That meant that the panel could operate even without purge, running the risk—likely a tiny risk, but a risk nonetheless—that in an emergency situation, it would touch off a fireball.

Williams said that he had worked on that system during the

last rig move and discovered how to make the automatic system work, keyed to the purge or lack of purge in the drill shack.

"Do you want me to start it back in automatic?" Williams asked.

"No," Williams says he was told. "The damn thing's been in bypass for five years. Why did you even mess with it?"

It sounded callous, but there were almost always two sides to every safety equation. Mike Williams wasn't wrong to worry. While the chance of the BOP panel igniting a fireball was remote, it was a real possibility. But a potentially more troubling possibility was that if the BOP panel shut down during a gas event, the driller would be left helpless, with no way to close off the well himself. It was a catch-22 that could only be resolved with the correct parts, parts that Williams knew had been on order for some time but had yet to arrive on the rig.

Jason Anderson would always tell anyone who'd listen how much he loved his work. But on this well, he was feeling the pressure. Doing things right, the way he'd taken such pride in learning, sometimes meant taking more time. The wear on the rig that went unattended and the mechanical breakdowns, combined with the exhortations from the BP men to hurry, were all starting to make him uncomfortable.

When this hitch ended and he went back home, he confided in his dad, Billy, a former high school football coach who'd gotten into the offshore business himself and steered Jason to his first rig job. Jason told his dad about the pressure to just get things done even if it meant cutting corners. In the past, he'd always been able to talk the company men out of something when he really felt it was important. This time, he told his dad, the pressure was more intense than ever, and he was worried he was losing the argument.

It was an odd coincidence, but even as his worries peaked, surveyors for a risk management company showed up on the Horizon, contracted by Transocean to conduct a confidential survey. The survey suggested that Jason's concern was shared by others. An analysis concluded that a significant number of workers worried that the quest to keep drilling always trumped the need for maintenance, forcing them to work with equipment that was becoming unsafe to use.

One man's comment to the surveyors summed up the general frustration. "At nine years old, Deepwater Horizon has never been in drydock," he said. "We can only work around so much."

Both Transocean and BP put a lot of money, time, and effort into promoting the "core value" that any worker at any time could stop work he deemed unsafe. But half the workers surveyed said they feared that if they spoke up, especially about things being controlled by managers in Houston, they'd face serious reprisal.

With that fear hanging in the back of your mind, it could be hard to speak up. It wasn't even just speaking up. What if you made your case, and the boss said, "I hear you, but we're going ahead and doing it my way," or the more common "I told the beach, it's in their hands now." How far were you going to go? The decisions were never black-and-white. Drilling a well was intensely complex and inherently risky. If you wanted to be 100 percent safe, you probably should never board a rig in the first place, nor start digging a hole in the ocean. And since risk was never entirely eliminated, you were never debating safety and danger in absolute terms. Millions of dollars were spent in accordance with percentages: How much was it worth to reduce the risk of a bad outcome from 1 percent to a half a percent?

After all, the Horizon had been drilling wells for a decade now, and nothing catastrophic had ever happened.

When his three weeks were up, Jason headed back to New Orleans and Port Fourchon, Louisiana, for that helicopter ride to the rig. At Macondo, things were starting to get interesting.

Up to this point, the hole had been passing mostly through sand and rock. The little pockets of gas it had hit along the way were mere soap bubbles compared to where they were headed—a fifty-foot-deep lake of liquid oil and natural gas permeating spongelike sandstone. Squeezed on all sides by billions of tons of rock exerting pressures of around 25,000 pounds per square inch, once pierced, it would explode to the surface as if a giant had stomped on a tube of toothpaste.

And the whole point of well construction was to build in such a way that the enormous force of upwelling oil and gas could be controlled, neatly shuffled into pipelines and sent harmlessly away to refineries. Job one was to ensure that this pipe they were sinking down, under enormous pressures deep inside the earth, did not leak. That would have been easier if a well were a single seamless cylinder of thick steel from top to bottom. But Macondo, and all deep wells, had to be made of pipes with seams and screw-together ends, and with smaller pipes hanging from larger ones—each connection a prime invitation to a leak. The first defense was to make sure that every individual section of the well was sealed up tight with cement.

Aboveground—pouring the foundation for a new house, say—cementing sounds simple enough. A plywood form is nailed together outlining the perimeter of the foundation, a cement truck backs up, dangling its cement chute over the mold, and pours down liquid cement until the form is filled. When the cement is fully cured, workers break off the plywood and the foundation is complete. What the homeowner does not see is more complicated. The humidity, ground temperature, and location all determine the amount and types of material—rocks, chemicals, silica—that go into the slurry to ensure the right drying time and strength of the hardened product.

Cementing a well is even more complex and much more difficult. You can't just pour cement in at the top and let gravity pull it down. You first have to get the correct mixture of chemical additives figured out, then find a way to pump the cement so that it flows out the bottom of the pipe and is forced up the thin annulus space between the pipe and the well wall, which is where the seal needs to be. As you pump the cement, it has to be untouched by any contaminants—like drilling mud or seawater, one or the other of which fills the hole at all times. Any mixing of those things with the cement will destroy its chemical integrity and render the cement job worthless. And you need to be certain, without being able to see it, that the cement went exactly where you meant it to go, and rose to an exact height in the annulus. To recap: The goal is to pump cement into spaces a couple of inches wide and completely submerged in water or mud without letting the water or mud contaminate the cement, using only tools that can be lowered through a narrow opening and operated from thousands of feet above.

It sounded like a trick even David Blaine wouldn't dare attempt. But the rig had its own magicians. They had nothing up their sleeves, but they did put a "shoe" on the bottom of the casing. The shoe had rounded edges that helped guide the casing as it was lowered to the bottom without scraping the sides of the hole or getting caught on ledges. A high-pressure nozzle that could pump mud or cement was lowered down into the casing pipe. To make sure that the pipe was clean and free of debris, mud was pumped through the pipe and out ports in the shoe. It hit the bare bottom of the hole, which forced it back up. The shoe, and the high-pressure flow coming out the holes in the shoe, prevented the mud from flowing back up the pipe. It had nowhere else to go but up the space between the outside of the casing and the bare wall of the hole until it came back out the top, carrying any debris with it.

After the mud had been circulated through the pipe, cleaning it out, the rig measured and pumped something called spacer, a water base with additives to make it friendly to cement. You could count on rig terminology to tell it plain: Spacer was there to put a space between the mud now filling the well and the cement soon to follow. It pushed the oil-based mud ahead of it up the annulus, readying the space for cement. A device called a plug was inserted into the pipe. The plug was like a cork within the tube. It had two parts, a top half and a bottom half, each of which had polymer edges that made a tight seal against the interior of the pipe.

When cement pumping began, an instrument called a "plug dart" was shot down a tube and hit a trigger in the top of the cement plug that unlocked the bottom half of the plug. The force of the cement flow through an opening in the top half of the plug pushed the bottom plug down the pipe. The plug in turn pushed the spacer ahead of it, out the holes in the shoe at the bottom of the pipe, where it made the U-turn at the well bottom and pushed up the annulus.

Back in Houston, engineers had run a computer model of the cement job, which had calculated the exact amount and chemical composition of cement and forcing pressure needed to ensure that the annulus was filled to the desired height and density. The cementer on the rig only needed to make sure the right amounts of additives were mixed in at the right times and count the number of pump strokes it took to push that amount of cement down the hole. When that number was reached on the stroke counter, another dart was sent down the pipe. This one sealed the hole in the top plug and released it to move down the pipe behind the full load of cement. Now they began to pump mud again. The high-pressure flow of the mud forced the top plug down behind the bottom plug, with the premeasured amount of still-uncontaminated cement between them. It was like a cement motorcade with armed guards in the front and the back.

As the bottom plug neared the shoe, it hit a barrier, called a float collar. The motorcade slammed to a halt. But as the pumping continued, the pressure increased to the point where a rubber membrane in the bottom plug ruptured, allowing the cement to flow through a valve that opened like a trapdoor in the float collar. The mud pressure continued to press down on the top plug, which pushed the remainder of the cement out the trapdoor and through the holes in the shoe. Just like the mud and spacer before it, the cement made the U-turn from the bottom and up the annulus, displacing the spacer.

Finally the top plug bumped down on the bottom plug, both snug against the float collar, and the pumping stopped. Gravity pulled at the cement now in the annulus, creating a slight backflow through the holes in the shoe, reclosing the trapdoor in the float collar. By measuring the "returns," the amount of mud that got pushed out of the well by the inflow of cement, the cementing team could see that it was the full amount, and therefore could assume that all the cement had been pushed where it was needed.

Now all they had to do was wait on the cement to harden. Even that had its own acronym: WOC, for Waiting on Cement. It was no joke. If they didn't wait long enough before they began to work the well again, the cement job, however perfect when it went in, could be ruined by stress before it had properly hardened. They just had to sit and wait, not the easiest thing to do on a rig, especially one so badly behind schedule.

CHAPTER TWELVE

A LONG STRING

Early April 2010

Macondo Prospect

As he drove to the airport to catch the flight to New Orleans, on his way back to the Horizon for the last time, Jason Anderson looked over at his wife and asked, *If something happened to me, would you and the kids stay in the house?*

He'd been in a strange mood these past few weeks at home in Midfield. You could have called it morbid. Shelley didn't get it. He should have been celebrating—his promotion had finally come through. He was going to be senior toolpusher, the person who directed the entire drilling operation on the Discoverer Spirit. His only superior on the drill floor would be the OIM. In fact, it was common for senior toolpushers to *become* an OIM. But instead of enjoying the news, he'd been worrying about things on the order of: Did Shelley know how to drain the RV water system and get it winterized? Would the kids learn the right way to handle a hunting rifle if he weren't around to teach them?

After years of putting it off, he sat down with Shelley and drew up a will.

Shelley had lots of experience running the house and tending to the kids for long stretches without a husband's support. She was used to it, still proud of the work Jason did. This time was tougher, though, thinking about all the things he'd been saying, and the fact that he shouldn't have been leaving in the first place.

The plan had been for Jason to spend time with Shelley and the kids before moving on to the Spirit. But the top Horizon managers had called to say they needed him on the Horizon, one last time, to help ensure that the new drill crew leadership was ready to take control. It was no surprise. Only three people were stationed in the driller's shack, the toolpusher, driller, and assistant driller, and like the Three Musketeers, the team would have to merge into a flawless unit when the well barked. His trinity was already missing a man. Barney Ray, the assistant driller, had been promoted to driller aboard the new semi-submersible Champion just the previous hitch. Jason had supported Ray's promotion and gave his thumbs-up to the selection of Roy Wyatt Kemp, a derrickman, as his replacement. Jason was confident in his own ability to watch over and mentor the new recruit. But the promotions, like dominos, meant the rest of the drill crew would also be promoted to fill the gap . . . the entire chain moved into jobs above their level of previous experience. This was not a time to replace the leader, especially on this job.

His previous hitch had been the worst yet. The new section of the well proved even more troublesome than the first. They had it listed on the well plan at six to ten days and three thousand feet. Instead, it took nineteen days and they got only half the depth. It had been a vicious cycle: feeling like they had to hurry, then pushing too hard and blowing holes in the well. They'd been putting so much sticky well patching material in there to stop the mud loss, that the drill pile got stuck. They spent forever, using every trick they knew, but they couldn't fish it out. The well services contractor Schlumberger came out to run sophisticated instruments down the drill pipe to determine where exactly the pipe was hung up. The measuring instruments got stuck, too, and they weren't cheap. All of it was costing BP, big time. They ended up having to do a sidetrack—to cut the pipe above the obstruction and drill a new path around it.

Through it all, Jason had felt uncomfortable with the pressure: the pressure readings down the hole and the pressure from the company men and from the beach to hurry up and get it done. By the end of his hitch, the date had come and gone when the Horizon had been scheduled to start another well. They were nowhere near the end of Macondo and had already overstayed their welcome.

The final section of the well gave neither Jason nor the Horizon any relief. The geology here was maddeningly inconsistent, and the mud weight needed to contain pressure in some areas was too much for others, pushing out the well wall and causing mud losses all the way down to the pay zone. They had to keep stopping to run in the well patching material, then going forward only to start losing mud again when the well walls gave way.

The primary objective was a 123-foot layer of sandstone, the top 53 feet of which was filled with hydrocarbons. Sandstone is most spectacularly familiar as the rock in the walls of the Grand Canyon. It looks solid enough, but in fact it is filled with microscopic pores. Under the high pressures of the deep earth, oil and liquefied natural gas can saturate it like a sponge.

They hit the primary oil deposit where the geologists had told them it would be. They knew they were there when the mud coming back up the riser and through the mud processing equipment began to be saturated with gas. But immediately below the deposit, as they began to drill toward the secondary objective, the mud almost stopped coming back at all. It was as if the bottom of the well had just dropped out—taking about three thousand barrels of mud with it. BP was now literally pouring money, about $1 million, into a bottomless pit. Even more money was tossed on top of what had already vanished, in the form of the well patching "lost circulation material," to seal the hole and stop the losses.

BP execs had had enough. Macondo had beaten them into an early retreat and it became clear that steel and cement were the only remedy. On April 9, they chose to stop drilling at 18,360 feet beneath the rig, or 13,293 feet below the wellhead. This was just far enough below the main deposit to allow them to do the cementing they needed to complete the well, but it was more than a thousand feet short of the secondary objective, another possible pay zone the Horizon had been meant to explore. It was no longer worth it. Everyone just wanted to get the hell out of Macondo.

All along, the primary purpose of this well had never been to begin siphoning the oil and shipping it to a refinery. From the start, the Macondo well was meant to determine just how lucrative the oil deposit would be. There had never really been any question of finding *no* oil. The seismographic studies had been unambiguous. The oil was there. The question was, how much, and more important, how quickly could it be extracted. Would the oil truly be worth the exceedingly high cost of pumping it out of the ground and into an oil refinery?

Building a major production platform to suck Macondo dry over a period of years—a rig that would be owned, crewed, and operated by BP—would cost a multiple of what R&B Falcon had spent on the Horizon. The BP flagship production unit, Thunder Horse PDQ, was built in 2004 for $5 *billion.* It was a semisubmersible like the Horizon, only with about 10,000 square feet more deck area and permanently moored rather than dynamically positioned. The design phase alone for a new platform would be six to twelve months, followed by another year or two to build it. The better part of yet another year would be required to ship it from Korea and install it in the Gulf with permanent moorings and connect it to the pipelines that snake all over the bottom of the Gulf, linking offshore wells directly to the shore-based refineries. The end product would be a multibillion-dollar octopus floating on the surface, its tentacles stretched into the mud deep below.

Bottom line: Macondo sat atop an oil field with real commercial potential, but after rushing so desperately and enduring so much adversity to complete the well, it would be cemented top and bottom, then abandoned for years before it would produce any oil.

The goal in completing the well was to make sure that it was (1) absolutely secure sitting untended for years, and (2) able to be transformed into an efficiently producing well, one that could be put on line fast and with minimum expense and trouble when the time came.

Like everything about this well, the completion promised to be difficult. The final section of the well was the portion most exposed to oil and gas deposits. So the fact that significant weakness in the formation had been encountered right at the bottom, causing 3,000 barrels of mud to escape into fractured rock, did not augur well.

BP's engineers were considering two plans to deal with the situation. One involved a two-step process. First, they would hang another stretch of pipe that would run from a hanger built into the end of the casing already in the well to the bottom of the hole.

When that was cemented in place, they would insert a steel tube connecting the top of the hanger back up to the wellhead. This was called a "liner tieback" and it was, finally, the tube through which the oil would actually flow when the well was producing.

The other option involved running a single string of steel casing from the bottom of the hole all the way to the wellhead—which was called a "long string."

The long string was BP's preference from the beginning. It was the simpler option. And less expensive. Back in March, a BP engineer named Brian Morel noted in an interoffice e-mail that the long casing string "saves a lot of time . . . at least 3 days." He followed that up with another e-mail to Sarah Dobbs, the BP completions engineer, and Mark Hafle, another BP drilling engineer, that "not running the tieback . . . saves a good deal of time/money."

That was important. The well was already more than $50 million over budget and it was the engineers' responsibility to find ways to stop the bleeding. They had a personal stake, too. BP employees were graded annually on how much money they save the company, under the evaluation category "Every Dollar Counts and Simplification." To prove their case, they compiled itemized lists of all the changes they instigated that reduced BP's costs.

This created a strong incentive to find the cheaper option, and the long string was definitely cheaper.

But it would also be riskier. With the single pipe going all the way to the top, any problems with the cement job at the bottom of the well would create an open path for oil and natural gas all the way to the wellhead, leaving the seal assembly at the top of the well—never designed to resist heavy pressure—as the only barrier, like a

160-pound placekicker left all alone to stop a 200-pound opponent from returning a kickoff for a touchdown.

The two-part option would create four barriers to gas flow. The first and most formidable barrier would be the cement job sealing off the bottom section of the well from the oil and gas deposit. If that failed, rising oil and gas would still have to break through a seal installed at the top of the final string of casing, on the hanger where it attached to the existing casing in the well. And that seal would be backstopped by yet another cement job right on top of it, where the "tieback casing" that ran all the way up the hole would be cemented at its base. Finally, there would be the relatively weaker seal at the wellhead.

Without those two extra barriers, only the cement at the bottom would stand between millions of gallons of oil and gas in the deposit and the vulnerable seals at the wellhead. Everything would be riding on achieving a quality cement job.

On the morning of April 15, that fact was making Jesse Gagliano sweat.

Gagliano was thirty-nine, a sharp-faced man with close-cropped brown hair that was graying at the temples. He was a technical sales advisor for the cement contractor Halliburton, but he had worked out of the BP offices for the four years in which he'd been assigned to assist BP with Deepwater Horizon projects. He'd been working on Macondo from the beginning, lending his expertise to the BP well team, designing the cement jobs and running tests and simulations to ensure that they would be successful.

He knew better than most that Macondo had been troublesome, in large part because of its geology. On the one hand, it had relatively high pore pressure—the accumulated weight of all the rock and ocean upon the hydrocarbons trapped inside the sandstone.

To keep the oil and gas from pushing out of the sandstone and up the well, the drillers had to put at least an equal amount of pressure down the hole, in the form of mud. That's why the drilling mud could be made heavier with additives if needed—the weight of the mud multiplied by the depth of the hole equaled the downward pressure available to hold back the upward pressure of the oil and gas trying to escape.

But there was another variable that made this simple equation complicated: Each section of rock a well encountered had a breaking point. There was just so much pressure the formation could take before it began to fracture. If it fractured, the well lost its integrity, and the mud would start pouring out of the well like water from a leaky cup. The fact that the pressure needed to keep the oil and gas down was very close to the pressure that would punch holes in the well walls was what made the bottom of Macondo so difficult. *Enough* pressure was perilously close to *too much* pressure.

In these circumstances, drilling was difficult. Attempting to get a successful cement job could be a nightmare. They were literally between a rock and a hard place.

As Gagliano began to build a cementing model for Macondo's bottom-to-top long string, all the particulars conspired against him.

Even the basics of cementing, like cleaning out the well first, could be risky in Macondo's case. Drilling mud tends to gel like fat in a refrigerator if it's been stagnant for long. Halliburton's "best practices" insisted on a full "bottoms-up" circulation of the mud, which meant that all the old mud would be pushed out and replaced with new mud, or at least thoroughly loosened up. This had the added advantage of pushing any trapped gas to the surface, where it could be safely disposed of. But in fragile Macondo, applying enough pump pressure to circulate bottoms-up risked pushing out a wall.

On the other hand, if the cleansing mud flow was too gentle or too brief, in deference to fragile well walls, thick clumps could be left behind, blocking the cement from filling the entire space and leaving channels through which dangerous amounts of oil and gas could rise unbidden through the well toward the surface.

Two factors idiosyncratic to the long string in Macondo made this last issue critical. One was that the space between the outside of the casing and the well wall was thinner than planned—a 7-inch pipe in an 8½-inch hole. BP had intended for the final segment of the well to be nearly 10 inches wide, but with all the problems they'd encountered, 8½ was the best they could do. They could have shrunk the pipe, but 7 inches was the minimum size that would make economic sense when it began to pump oil to refineries.

The problem for Gagliano was that the thin space added to the difficulty of forcing the cement to flow evenly and completely around the outside of the pipe. If the pipe weren't perfectly centered in the hole, it would be kissing up so closely against the side that no cement would go there at all, choosing instead to flow in the path of least resistance, on the fat side of the space. This was called "channeling," and it meant that there would be little voids in the cement, allowing hydrocarbons to sneak through the annulus to the top.

Making matters worse was the fact that the MMS required them to seal the annulus with cement to at least 500 feet above the highest oil or gas deposit. The top of the primary deposit was 277 feet above the hole bottom, but a thin hydrocarbon zone had been discovered another 262 feet above that, meaning the cement would have to extend at least 1,039 feet up the annulus. They had no choice but to pump heavy cement—heavier than the mud—up a long, thin space in a fragile hole. The only way to make that happen was to push down more cement, harder, and for a longer time. Given Macondo's tendency to crumble, it was a pretty challenging scenario.

Too challenging. When the computer spit up the results of the simulation, its judgment was harsh. The notation on the report read: "unlikely to be a successful cement job due to formation breakdown."

Gagliano knew he had bad news in his hands of the urgent variety. Fortunately, he was only feet from the BP drilling engineers working on Macondo; his office was on the same floor of the same building on BP's Westlake campus, west of downtown Houston. He printed out the report and strode out to show it to them. As he turned the corner in the hallway, there they were: Brett Cocales and Mark Hafle.

"Hey, I think we have a potential problem here," he told them. And he showed them the printout, pointing out where it indicated the potential for gas flow from the deposit and channeling in the cement.

The BP engineering team huddled and quickly came up with a new analysis that cited the unfavorable cement modeling results and recommended that they now go with the previously discarded two-part tieback option because it provided multiple barriers against cement failure.

Within BP, the recommendation was rewritten yet again in a matter of hours, flipping the conclusion. In a well that had already gone so far over budget, the casing hanger and tieback option would "add an additional $7–$10 MM to the completion cost," the new recommendation said.

It also noted that the same thing that made a long string vulnerable—it was a straight shot from bottom to top—would also make it more durable and less prone to leakage during long-term production use.

Gagliano was sent back to the drawing board to find a way to hold on to the long string. He and Cocales and another BP engineer, Greg Walz, worked into the evening to find a solution. They had already decided to use a special cement impregnated with nitrogen bubbles—it had a texture reminiscent of shaving cream—because it was considerably lighter than normal cement and wouldn't put as much stress on the well when it was pumped in. Plus, the nitrogen bubbles stiffened the wet cement and made it more resistant to disappearing into a fractured formation—similar to how shaving cream sits in a sink, refusing to go down the drain. But foamed cement alone didn't make the model work. Now they focused on the danger of channeling, finally realizing that it could be minimized if they could ensure that the pipe would remain almost exactly in the center of the hole from top to bottom. This would eliminate the stray sections where it kissed up against the wall and obstructed the cement from rising.

There were simple devices that could accomplish this. This was the oil industry, so of course centering devices would be called "centralizers." They were like metal springs that fit around the pipe, projecting equally in all directions. As the pipe was lowered into the hole, the centralizers would push against the walls, guiding the pipe into the exact center of the hole. Gagliano's initial model had used some centralizers, but by experimenting, they discovered he just hadn't used enough. Finally they discovered that twenty-one centralizers spaced along the pipe would bring the model to an acceptably small risk of cement failure. And just to be sure, they recommended the use of an expensive testing procedure called a cement bond log, in which a sophisticated sensor would be sent down in order to determine if the cement had formed a reliable seal with the exterior of the casing.

The cement bond log test would take nine to twelve hours of rig time and cost $128,000 on top of that. But if any channels had been left, it would detect them. The channeling was a serious problem, but correctable—it would take another time-consuming and expensive procedure, but it could be done. The casing in the troubled areas would be perforated and new cement would be pumped through the holes using a special tool. Then those sections would be recased.

If the engineers were pleased with themselves for coming up with a solution, it didn't last long. When Brian Morel, the BP engineer who'd been dispatched to the Horizon to work on site with the rig's BP company men, Sepulvado and Don Vidrine, heard about the idea of increasing the centralizers, he e-mailed back: "We have 6 centralizers, we can run them in a row, spread out, or any combination of the two. It's a vertical hole, so hopefully the pipe stays centralized due to gravity. As far as changes, it's too late to get any more product on the rig, our only option is to rearrange placement of these centralizers."

They all understood what the "too late" part of that message meant. Morel apparently wasn't about to advocate delaying the drilling of Macondo one more time just to wait on a handful of metal rings. They also understood the severe consequences if the computer prediction came true. It wasn't his job—he was a well engineer, not a parts manager—but Walz got on the phone and not only hunted down the fifteen additional centralizers, but arranged for them to be flown out to the rig on an already scheduled flight the next morning along with personnel to install them. Walz sent a memo to his boss, BP's Macondo well team leader, John Guide, telling him what he'd done, adding, "We need to honor the modeling to be consistent with our previous decisions to go with the long string."

Walz must have been worried that he was overstepping his bounds, and feeling defensive about asking for changes in a cement job so late in the process. In the next sentence he cowered a bit before the boss: "I do not like or want to disrupt your operations. . . . I know the planning has been lagging behind the operations and I have to turn that around."

His attempt at diplomacy didn't appear to have much effect. Guide messaged back that he thought the centralizers had a kind of fitting that could fall off in the hole and cause a whole set of new problems. It turned out Guide was mistaken about that, but he had other reasons to turn down the centralizers Walz had rustled up. "It will take 10 hrs to install them," he wrote. "I do not like this and . . . I [am] very concerned about using them."

Guide's word turned out to be law.

Brett Cocales, one of the engineers who had worked into the night with Gagliano to come up with the twenty-one-centralizer solution, couldn't completely mask his bitterness at the outcome. That same day, he e-mailed Morel: "Even if the hole is perfectly straight, a straight piece of pipe even in tension will not seek the perfect center of the hole unless it has something to centralize it."

In what comes next, you could almost see him briskly brushing his hands:

"But, who cares, it's done, end of story, will probably be fine and we'll get a good cement job."

BP applied to MMS for an amended permit to run the long string. The application was approved the same day.

As BP engineers in Houston wrestled with how to finish off Macondo's exploration phase, a helicopter full of men in coveralls with the logo of ModuSpec, a risk assessment contractor, descended on the Horizon and began climbing and crawling over every part of it, bow to stern. In 2011, the Horizon was scheduled to go into drydock for the first time since it was launched in Korea. Transocean wanted an independent assessment of what needed tending. When the ModuSpec crew finished crawling through tight spaces and marking up their clipboards, they spent days poring over paperwork and interviewing crew members. The audit was completed on April 14. It identified hundreds of items of urgently needed maintenance, ranging from repairing an out-of-order thruster to complete rewiring of the pipe-racking system. It also noted this: The blowout preventer was significantly overdue for its five-year inspection.

CHAPTER THIRTEEN

UNEASY PARTINGS

April 14, 2010
Sandtown, Mississippi
Ivoryton, Connecticut

On Wednesday night, the night before Dale Burkeen, the Horizon's starboard crane operator, left for his late April hitch on the Horizon, Janet Woodson told her husband she wanted to go see Dale off.

Janet, Dale's little sister by eighteen months, hadn't seen much of her brother this time, and she wanted to make up for that. Dale had spent his first few days off, as he usually did, just staying at home in his tiny central Mississippi community of Sandtown with Rhonda, the wife he adored, and six-year-old Timmy, whom he always called Bo.

Janet didn't want to interrupt. He and Rhonda hadn't seen each other in two weeks, after all. Dale was sometimes so gooey with love, even around company, sweety-pie-ing and baby-ing her, that Rhonda would get embarrassed and have to tell him to shut up. "Isn't she just the purtiest little thing?" he'd say.

And that little boy meant the world to him. Dale would take

him just about everywhere and was always buying him things. She heard he'd just gone out and bought the boy a rifle with a scope in honor of his upcoming seventh birthday, relishing the idea that this would be the year Bo brought down his first buck.

Dale was a big playful boy himself—*very* big at six foot one, 280 pounds. Easter Sunday, he'd thrown a barbecue for the whole family, grilled up a mountain of burgers for their parents and all the cousins, nieces and nephews. But before they ate, Dale had insisted on an egg hunt. Only he didn't want to be the one to hide the eggs—he wanted to hunt them with the kids. Classic Dale.

Janet had missed out on all that. She'd been camping in the mountains with her husband. That's part of why she was driving out to Sandtown on the evening of April 14, to make up for lost time.

She rolled through pine thickets down the familiar honeycomb of gravel roads to his double-wide trailer home with the brown siding, in the middle of nowhere. She and Dale stood out by the road in Dale's front yard at dusk for about an hour and a half, talking as night fell. Two bright security lights on wooden poles that flanked the little house came on, reflecting off the sandy, patchy front yard and the road in front. Dale never wanted to give up his last day at home, so instead of driving down to the coast at a reasonable hour and getting a hotel room, he would leave around midnight and arrive at the helipad for the morning ride out to the rig. Janet always fretted about that part of the journey. She worried about Dale making the drive when he was sleepy. And she worried about that helicopter ride to a tiny target in the middle of that big ocean.

Sometimes she wished he didn't have to work so far away, and stay away so long when he was gone. But she understood, of course. Everybody in rural Mississippi and Louisiana understood.

Out in front of the house, there was a little plastic wading pool sitting next to a leaky faucet. Little fish were painted on it, swimming tentatively in airy circles above a fingernail's worth of water in the bottom. Just the other day Dale told Janet, "Sis, I'm going to get a real swimming pool for all our kids and nieces and nephews and other kids in the neighborhood."

She knew he'd do it, too.

But it wasn't just the money that kept him on the rig. Dale enjoyed being out on the water for that long stretch of days. He talked all the time about how pretty the Gulf was and how he would see dolphins jumping and fishermen catching big fish out by the rig.

Janet loved that about her brother, how he could find so much to appreciate in the world. And she loved that he was such a good listener. She always felt safe around him, not just because of his size and his fierce protectiveness, but because she knew he always had her best interests at heart.

Janet felt she could tell him just about anything. Now she found herself telling her brother about her concern for her son, who was about to go into the sixth grade. Janet was a special education teacher, and she knew all about vulnerabilities in children, maybe too much. She was worried that he might be too young for middle school, wondered if she should hold him back.

Dale put his big bear arm around her and gave her a smile so full of love it made her shiver. He said, "Oh, sis, Scott's gonna be okay, and you're a good mother and don't let anybody tell you any different."

Janet cried, and he hugged her and gave her a kiss as they stood there in front of the house by her car parked on the road.

The last thing she said to him was, "Be careful."

Alyssa Young had been through it dozens of times. She knew all the feelings. There was the heightened appreciation of Dave's *presence*—the rhythm of his breathing beside her in the dark, the laugh that turned everything into their private joke, the way he could lose himself in play with their children, almost childlike himself. And then there was the creeping tension that accumulated as his departure day neared, and the battle against her own self-pity for being left alone to deal with three small children and a needy dog in a big house at the end of a lonely New England gravel road.

She reminded herself she had signed up for even worse when she married Dave. He was out on that cable repair ship for three or four months at a time, barely reachable. Since he'd been on the Horizon, she spoke to him practically every day, at least once, usually more. And three weeks was a far cry from three months. Just twenty-one days until they'd have three blessed and almost completely unobstructed weeks together as a family.

She knew she was luckier than so many others. But the void when he left, and the anticipation of it, never seemed to get any easier. And every time the day came for her to drive him to the airport to catch his flight to New Orleans, all the complex calculus of gain and loss would swirl through her mind as another good-bye rushed toward her at sixty miles an hour.

But this time there was something else in the car with them, something she'd never felt before—a sheer animal dread, a suffocating presence that clutched at her and wouldn't let go. This was nothing like the usual knot of anxiety that she'd come to associate with the helicopter flight from Houma, Louisiana, to the rig. That was at least a partially rational fear. She'd looked up the stats and knew that he was safer on the commercial flight to New Orleans than they both were on the drive to the airport. But helicopters were something else. She knew all about the ones that had gone down while shuttling workers to rigs, and she always half held her breath until he called to say he'd touched down on the Horizon. Then she didn't think about it again until he was on the helicopter back to shore.

That familiar, nagging twitch of concern had no relation to what gripped her now. She came close to telling Dave what she was feeling and begging him not to get on the plane, but that would have been crazy, she knew. She didn't believe in this. She refused to believe in it.

The sick feeling persisted. It stayed with her on the lonely drive home, swirling around her in a black fog until Dave called the next day, April 15, to say the helicopter had landed on the rig. She could feel her muscles relax. Now he was safe.

CHAPTER FOURTEEN

POSITIVE TEST

April 19, 2010

Block 252, Mississippi Canyon, Gulf of Mexico

Jimmy Harrell didn't like what he was hearing.

He was at the 11 a.m. meeting on April 19. They had been running the final long string of casing down into the hole since the day before, and now they were looking ahead to the final significant task before leaving Macondo behind—the cement job that would seal off the bottom of the well. Ronald Sepulvado, the senior BP company man he had worked with for years, had been called to shore to participate in a mandatory well control school to refresh his knowledge of how to deal with a well that threatened to go out of control. Sepulvado's temporary replacement was Bob Kaluza, a man who'd been flown to the Horizon from BP's biggest asset, the Thunder Horse PDQ production rig.

Kaluza had plenty of experience, thirty-five years in the oil field, but having spent just a few days aboard the Horizon, Sepulvado's notes could not have adequately conveyed Transocean culture, the personalities of the crew, the specifics of the equipment the Horizon used, or how to handle operations on an asset not owned by BP. But here he was nonetheless, outlining to the Horizon's drill crew the plan for the upcoming cement job.

Something he said took Jimmy by surprise, and taking Jimmy by surprise on a rig was not an easy thing to do. Rigs had been his life for more than three decades, and he'd worked his way though just about every job a man could have on the rig, beginning with deckhand, and he'd seen just about everything there is to see.

Now he was the OIM, the boss, and the senior drilling hand aboard. Like so many in the Gulf, he was a southerner, born in Mississippi, and a true gentleman, careful to call the men in positions above him, even men years his junior, "Sir," and all women, regardless of age or position, "Ma'am." His round face, soulful eyes, and droopy white mustache made him seem approachable, like a favorite uncle. Everyone just called Jimmy "Jimmy," or "Mr. Jimmy" if they were a junior hand and felt presumptuous addressing the boss by his naked Christian name. But Jimmy wouldn't have minded if they had. He had the kind of self-confidence that expressed itself as extreme calm in difficult situations. He had been in enough tight squeezes in his life to know that if you kept an even keel, tough times usually ended up for the best.

He had needed that faith on a well like Macondo, which had been a true bitch of a project right up to the end. Now this cement job would all but finish it up.

What caught his attention now, and kind of shocked him, was the company man saying they intended to run special, and more expensive, nitrogen foam cement for this job.

Let me stop you right there, Jimmy said. He knew a thing or two about nitrified cement. He was no stranger to the stuff; they used it all the time—they'd used it earlier on Macondo. But they'd always used it up near the top of the well, at the muddy sea bottom.

In his more than three decades in the business, he'd only seen it used in a deep part of the well twice.

The problem was that at great depths—and this cement was going down three and a half miles—the high pressure threatened to squeeze the nitrogen bubbles right out of the mix, which would then rise up the well, creating tiny holes and channels that would render the cement job worthless as a seal against oil and gas. Jimmy didn't like this at all, and he didn't even *know* about the tests Halliburton had been running on the proposed cement slurry—so far, the mix had failed to prove itself stable.

His objections came from the gut, and his long experience. But Kaluza wasn't giving in. Lighter cement, less likely to break through the well's fragile walls, was essential to the plan. The nitrogen had to stay.

Jimmy figured he wasn't going to win this argument. The company man was the customer, the man who wrote the checks. This was *BP's* project. But it was Jimmy's rig, and he wasn't going to let it drop without at least a little razor-edged sarcasm.

"Well," Jimmy said, "I guess that's what we got those pinchers for."

The cement job began at 8 p.m. on April 19. The first step was to make sure the hole was clean and safe and the casing ready to adhere to the cement. For this particularly delicate job, the standard practice called for a full bottoms-up circulation, achieved by pumping enough clean well fluid down the hole to push all the existing mud out the top.

This can be critical to getting a good cement job. Drilling mud's tendency to gel when sitting idle is a useful property. If it remained entirely liquid, whenever the pumps stopped during drilling, all the cuttings and debris the mud had been pushing out of the hole would just sink back down to the bottom. The gelled state is thick enough to hold the debris in suspension until pumping resumes. But that very property can introduce serious problems. Vigorous circulation of clean mud will break down the clots and push them out of the well. If the circulation is too gentle or too short, some of the clots will remain clinging to walls, especially where the casing is squeezed against the wall of the well—exactly the scenario Jesse Gagliano had warned about.

Incoming cement seeks the path of least resistance. It will flow around the clots rather than displace them, or force the clots to finally dissolve, only to contaminate the cement, riddling it with channels that can be invaded by pressurized oil and gas.

But BP elected not to do a bottoms-up. Doing a full circulation may have threatened to fracture Macondo's fragile walls. The pressure of the pumping had been reduced to avoid that, but reduced pump pressure also meant that doing bottoms-up would take hours of rig time—possibly more than BP was willing to budget.

After circulating only 342 barrels of mud, about half the amount that would be needed to do a bottoms-up, Halliburton contractors began to pump the cement.

They actually began by pumping a lighter-than-water base oil. This was yet another way to address the fear that the cement job would push out the well's bottom again. Even the lighter foamed cement was more than twice as heavy as the base oil. Stacked up in the annulus for 1,000 feet, the cement, plus 12,000 feet of drilling mud on top of it, would have added up to tons of pressure bearing down. Pumping in the base oil would displace an equal volume of drilling mud out the top of the riser, significantly lightening the overall load on the well's bottom.

After the base oil went in, followed by spacer fluid to prevent contamination from the oil, they began to pump in the cement itself. The amount had been carefully calculated and measured so it would fill 190 feet in the bottom of the final casing, then get pushed up, as gently as they dared, from the bottom of the hole into the annulus, reaching a thousand feet toward the surface. When all the cement was in place, they checked the returns in the mud pits. The drill crew carefully calculated the amount of mud displaced. It was a match for the amount of cement that went in—what they called "full returns." This meant that all the measures they had taken had succeeded in preventing the well walls from giving way.

But it didn't say much, if anything, about whether cement had formed channels, as predicted by the model. The cement could have channeled like crazy or failed to adhere to the casing and still give full returns.

The only way to determine whether the cement was solidly in place would be to let the Schlumberger crew that had been waiting on board do the cement bond log test. They would run their delicate instruments down into the well, dangling from a hook at the end of a wire cable, unspooling from a huge reel and making thermal and sonic measurements as they went. Computer analysis of the results would create a 360-degree image of the cement in the annulus, revealing any voids the cement had failed to fill.

But the BP engineers, both on the rig and in Houston, saw the full returns and declared the cement job a success. The Schlumberger crew would be sent home on a morning BP flight off the rig without conducting the bond log test, saving BP $118,000 and the half day it would have taken to run the test—amounting to roughly another half million dollars saved.

A few days earlier, the nitrogen foamed cement had been tested in a lab to determine how long it would take to set under the anticipated temperature and pressure conditions that would exist at the well bottom. The lab test revealed that after twenty-four hours, the foamed cement still hadn't set. It took forty-eight hours to reach maximum hardness.

But just short of eleven hours after pumping liquid cement into the bottom of the well, BP officials ordered what is called a "positive pressure test" to establish the integrity of their new long string casing. They closed off the well at the BOP in preparation, then pumped mud down the "kill line," which skirts around the annular preventer and can increase pressure at the bottom of the well even when the annular is closed. They kept pumping until 2,500 pounds per square inch of pressure built up inside the newly cemented casing. They held the pressure at that level for thirty minutes. If the pressure eased in that time, it would mean there were leaks.

The pressure held steady for the full half hour, and the test was deemed a success.

The fact that the positive test was conducted when the cement may not have yet been fully hardened could conceivably have caused a new problem. Pressuring up the casing makes the steel expand slightly. If the cement mix was still semifluid, the expanding casing could have broken the bond between steel and cement, leaving a small but potentially lethal space that gas could pass through on the way to the surface.

The cement job and the testing had gone on through the night, into the morning.

It was Tuesday, April 20, 2010.

CHAPTER FIFTEEN

NEGATIVE TEST

0630 Hours, April 20, 2010

Block 252, Mississippi Canyon, Gulf of Mexico

Miles "Randy" Ezell, the Horizon's senior toolpusher for four years and one of the original crew who brought the rig out of Korea, had his first meeting just as the sun broke the wavering curve of the Gulf. It was 6:30 and uncommonly calm with the promise of a beautiful subtropical spring day. The first meeting was a teleconference review of the coming day for Transocean management. That meeting rolled into a meeting he attended with the BP company men, conferencing in with their people in Houston to discuss the plan for the day, followed by the Horizon supervisors' meeting with all the department heads.

The department head meeting got going around 8:30, and it lasted for some time, mostly as they discussed the DROPS program, which was designed to curtail injuries due to one of the bugaboos of rig life—dropped objects. Nobody needed to tell them about the dangers of gravity in such a compact and intense industrial zone as the Horizon. Anything dropped off the 240-foot-tall derrick—

a bolt, a hammer, even a pencil—could became a deadly missile. And a load of steel casing dropped overboard during transfer from a workboat might just take out the riser or even damage the BOP, a multimillion-dollar mistake. Transocean attacked the problem in typical fashion—with an acronym, a set of principles, and an aggressive program of proselytizing to drive those principles home. This included signs, literature, tons of required documentation and forms to fill out, and, of course, meetings. The Horizon had a record and a reputation to protect. The rig had kept serious accidents to an admirable minimum. This very day, four VIPs from Transocean and BP were flying in to deliver an award to the Horizon for controlling the number of "lost time" incidents, defined as any accident resulting in a worker missing time from work beyond the day or shift the accident occurred. In fact, there had been no lost-time incidents on the rig for seven years. And as everyone in the offshore industry knew so well, lost time was lost money.

When the DROPS meeting was finally over, Randy had just enough time to make rounds of the rig floor before yet another meeting, this one the regular "pre-tour safety meeting," essentially a huddle for the oncoming crew to discuss plans for the day, led by Jimmy Harrell, the driller Dewey Revette, Randy himself, and some others. He walked into the meeting room near the galley as Dewey outlined the order of battle.

Today was the day they would begin closing up the well for "temporary abandonment," until a production platform was designed and built and installed above it. They had just put in place a cement plug at the bottom of the well. Now all they had to do was set another cement plug near the top. Federal regulations require that "top plugs" be installed no more than a thousand feet below the seafloor. But BP had received approval to put Macondo's top plug deeper than that, 3,300 feet below the mud line. The given rationale was to avoid disturbing the area around the wellhead seal. The unusual depth had an additional advantage: All the very costly mud above the top plug could be pumped to a workboat for transport to another BP project, instead of being left behind.

But not until the well was completely sealed.

Ever since the well had penetrated into the oil and gas deposit, the weight of all that mud had been keeping the hydrocarbons from shooting to the surface. Once the drilling mud was replaced with much lighter seawater, control of the oil and gas would depend on the integrity of the well's seals and plugs.

Dewey outlined the final test they would perform to ensure that Macondo was safe, and ready for mud displacement. The procedure was called a "negative test." In the "positive test" that had been run that morning, pressure had been inserted inside the well casing to see if anything leaked out. The negative test would take pressure off the well, to see if anything leaked *in*. The test was somewhat complicated and would take some time.

The consumption of time was apparently a sticking point. The first plan that BP company man Bob Kaluza presented for the day consisted of displacing the mud in the riser with seawater and setting the top plug. A negative test was not in the plan.

Jimmy and Randy had some words with Bob. Skipping the negative test "is not my policy," Jimmy said. He alluded to a bad experience he'd had years ago, but he didn't go into detail. "It taught me a lesson," he said. On the Horizon, Jimmy insisted, a negative test was standard until he heard differently.

Well, I'm the company man, Kaluza told him in effect, and you're hearing it from me. If the well passed the positive test, he said, the negative test was overkill.

Jimmy was adamant. He'd given in on the nitrogen foamed cement. He wasn't giving in this time. They were running that test.

The helicopter carrying the VIPs arrived on the rig at 2:30. It was one of those wave-the-flag, check-in-with-the-little-people management exercises. Two senior execs from Transocean, Daun Winslow and Buddy Trahan, had joined up with two BP counterparts, David Sims and Pat O'Bryan. They met in Houma and got lunch together before the 12:30 check-in for the helicopter shuttle out to the rig.

Daun was the ranking Transocean exec, officially the operations manager, performance, for the North American Division. Like many top Transocean officials he'd come up from the rig floor—he'd been an OIM for a decade, then stepped up to be a shore-based rig manager, the OIM's boss, and now he oversaw a vast swath of Transocean's operations. His job description was to oversee "safe and efficient operation, maintenance and equipment on rigs in Gulf of Mexico"—all without a four-year college degree to his name. His highest academic attainment was technical college and an auto body apprenticeship. He had a slightly unkempt look, thinning brown hair swept back like it had frozen in a strong wind. His lean build and smoker's pallor was underlined by physical confidence that conveyed the habit of command.

Buddy Trahan, forty-three, was the rig veteran who had flown in to help set the Horizon right after its near capsizing in 2008. He'd come from a large family that had run shrimp boats in the Gulf, and now he supervised six Gulf oil rigs worth more than an entire fleet of shrimp boats.

Together, Daun and Buddy hoped to use the visit to enhance the relationship with their biggest client, wave the flag, and show

O'Bryan, BP's vice president for drilling and completions, and Sims, the company's drilling operations manager, the in-the-field spirit of one of their most exemplary rigs.

These VIP trips had another purpose as well. The people who ran both BP and Transocean understood that an us-versus-them dynamic would inevitably arise between the workers on the rigs and the folks on "the beach," as the Houston headquarters were called. Mostly, the rig tours were an attempt to close the gap, at least a little. For those making the big decisions it was important not to lose touch, to see with their own eyes and hear with their own ears how their policies translated into the tough and dangerous work of pulling oil from the Gulf.

Even so, it wasn't in them to pass up a chance to proselytize. They went over a few points they wanted to address—the hazard of dropped objects, of course. Also, and they wrote this down, "Slips, Trips, Falls."

They saw a chance to reinforce the values that had served the Horizon so well. The "No blame, 'can do' culture—fix the problem, learn, move on." Also, "prudent risk-taking—freedom to fail, no fear of second guessing."

The intense concentration of Transocean and BP on relatively minor "slips, trips, and falls" had struck some as odd. They were, after all, sitting directly above a cavern of highly pressurized, highly flammable material that could erupt, with horrific consequences. But few questioned the motives of executives who were, after all, looking out for their safety.

The VIPs arrived at 2:30 p.m., signed in, and were handed the standard issue of rig life: hard hats, gloves, safety glasses, and earplugs.

BP's O'Bryan had never been on the Horizon before, so had to sit through a one-hour safety briefing. Transocean's Winslow, who'd had the briefing before, decided to sit in on it again—he wanted to see if the medic who conducted it had picked up on any of the "improvement opportunities" Daun had given him on his previous visit.

At the end, all four executives were assigned a lifeboat—a formality, obviously, but it was a requirement, even though they were going to be on the Horizon for less than forty-eight hours.

Throughout drilling, the well had been maintained in a condition called "overbalanced."

This meant that the pressure going down the well, in the form of the weight of the fluids that filled it, was greater than whatever pressure was pushing up from the formation toward the surface. The cement job that had just been completed should have meant that the formation pressure was now sealed off at the bottom of the well. Another seal, this one at the wellhead, had also been installed and pressure-tested that morning. The wellhead seal was a second barrier in case the seal at the bottom failed, but it wasn't designed to stop a high-pressure event, like a blowout. So it was critical that the cement job at the bottom, the bottom plug, be secure.

This is what the negative test Jimmy Harrell had insisted on would confirm. Before they removed *all* the mud that had kept the well safe through the drilling process, they would replace some of it with lighter seawater. Now the well would be "underbalanced"—the downward pressure caused by the weight of fluids in the well would be less than the pressure pushing up from the formation. If the cement job had worked, even in an underbalanced state there would be no pressure coming into the well.

To save time, Kaluza and the day toolpusher, Wyman Wheeler, decided the negative test would be done in lockstep with the other chores that remained before hauling the riser and BOP back up on deck and leaving Macondo behind. In one drill string's trip down the well—important because just getting everything down there took hours—they would do the negative test, replace the mud remaining in the riser by seawater, then install the top cement plug regulations required before temporary abandonment.

Wyman, forty, lived in Monterey, a tiny Louisiana town with no stoplights, no pizza delivery, one school that ran all the way from pre-K through twelfth grade, but more than half a dozen churches. Wyman and his wife Rebecca's eight- and ten-year-olds attended the town school, as had Wyman a quarter century earlier. Instead of going on to college he went right to work, ending up in the oil fields, and rising to toolpusher on the Horizon.

Wyman was thoughtful and cautious in his approach. He appreciated the complexity of testing in a well where the only information on what was happening thousands of feet down a black hole had to be deduced from meters, dials, and gauges on the rig.

To begin the test, the crew lowered the stinger tube into the well to the 3,300-foot depth where they intended to set the top cement plug. The stinger would start spraying spacer, the thick fluid used to prevent the mud from mixing with the seawater that was to follow. As it sprayed into the well, the mud below it compressed, having nowhere to go, and forced the spacer to make a U-turn and head back toward the surface.

The spacer was gray, dense, and sticky. And there was a lot of it.

It was actually leftovers from the long war against the crumbling walls of the well: two 200-barrel batches of different types of "lost circulation material," known, like everything else in the industry, by an acronym. LCM was the thick and pulpy patching material they'd needed from the start in Macondo. When they had finally stopped the bleeding at the bottom of the hole a few days earlier, these two batches—charmingly trade-named Form-A-Set and Form-A-Squeeze—were all that was left.

Leo Lindner, the Horizon's drilling fluids specialist, had been told by the BP company man to mix the two together for use as a spacer for the negative test. He'd never seen LCM used that way, and it was more than they needed, about twice as much. But Lindner could understand why BP wanted to use it. Any LCM left over would have to be shipped back to shore. With the workboat filling up on all the displaced mud, another service boat would have to come out to haul it away—more time and money. On the other hand, if it went into the well as spacer, when it came out, federal regulations said it could simply be dumped overboard.

But Lindner was still unsure about mixing Form-A-Set and Form-A-Squeeze together. Monday night, he poured a gallon of one into a gallon of the other to see if anything strange happened. Nothing really did. So the next day, the 454 barrels of the Set-Squeeze mix was fed into the stinger. As planned, the spacer U-turned and went back up the well, pushing the regular mud ahead of it, out the top of the riser, into the mud pits, from where it was eventually offloaded in a thick hose to the 260-foot-long workboat, the *Damon B. Bankston*, which had tied up to the side of the rig that morning.

After all the spacer was in the well, the stinger began pumping seawater behind it. Wheeler and his team planned to pump enough water to push the spacer above the BOP—just barely, by twelve feet. When they believed that had been accomplished, they closed the BOP's annular preventer—the rubber doughnut that sealed around the drill pipe.

Theoretically, this would hold all the spacer and mud above the test area, leaving only the weight of water against the upward pressure of the well. If the cement job had sealed properly, the inflow pressure from the formation should be zero, as should the pressure on the drill pipe.

But that's not what happened.

With the annular closed, the drill pipe pressure declined to 273 pounds per square inch, but no further. As the team puzzled over this, they discovered that the mud level in the riser had dropped about fifty barrels, which shouldn't have happened. That could only mean that the annular preventer had leaked. Nobody mentioned the handfuls of rubber that had been found in the outflow from the well recently. In any case, for whatever reason, some 2,000 gallons of the Form-A-Set and Form-A-Squeeze combination had dropped through the annular into the seawater test area, contaminating it.

At this point, the test was ruined. Whatever pressure readings they got would have to be interpreted through the noise of the mud that had leaked in. The situation could be rectified, but it would mean starting over, refilling the well with mud, new spacer, and new seawater to reinstate pristine test conditions. Then the annular preventer would need to be resealed, this time tightly enough to make sure nothing could leak back down.

But only the second part of that happened. The pressure on the annular preventer was increased from 1,200 pounds per square inch to 1,900 pounds per square inch, and the drill pipe line and kill line—the line that bypassed the annular preventer and connected below it into the well—were closed while they all discussed what to do next.

As they talked, the drill pipe pressure shot up to 1,250 pounds per square inch in six minutes.

Jimmy Harrell and Randy Ezell ushered the VIPs around the rig on their tour. It became clear that the subject they were most interested in was accident prevention. They had some recent incidents from other rigs in mind, and wanted to see if "lessons learned" from those could apply to the Horizon. They climbed out on the rig floor to see a piece of equipment where, on another rig, a worker had stepped into a twenty-four-inch depression, slipped, fallen, and dislocated a shoulder. They seemed a little disappointed to discover that the apparatus was entirely different on the Horizon. None of the learned lessons could apply.

Now they asked to go down into the columns, where the Horizon had had the flooding incident in 2008. The BP VIPs, Pat O'Bryan and David Sims, wanted to see what it looked like down there. Daun Winslow and Buddy Trahan asked if they could wander around, put eyes on the area and get a feel for what that had all been about.

First, though, the VIPs stopped into the driller's shack. It was standing room only in there. Something obviously intense was happening. Daun sidled through the crowd and tapped the driller, Dewey Revette, on the shoulder.

"Hey, how's it going, Dewey?" Daun asked. "You got everything under control here?"

"Yes, sir," Dewey said.

Daun sensed something wasn't quite right. He pulled Jimmy and Randy aside. They were the most senior Horizon hands aboard, and monopolizing them for the VIP tour suddenly seemed like a low priority.

"Looks like they're having a discussion here. Maybe you could give them some assistance," Daun suggested.

So Daun and the other execs went off to look at plumbing in the columns while their tour guides stayed behind to deal with whatever was going on in the driller's shack.

The first thing the drill shack assemblage did after realizing what had happened was to call over to the workboat, the *Bankston*, and tell its captain, Alwin Landry, that they were shutting down the mud transfer for a while. Landry assumed they were breaking for dinner. But in the drill shack, dinner was the last thing they were worrying about. The debate continued for about an hour. Wyman was convinced the pressure readings were evidence of a problem in the well. Bob Kaluza, the day shift company man, thought otherwise. He argued that the pressure spike on the drill string had been caused by the weight of the mud dropping back down into the test area.

It was past five, and the night shift guys—Jason Anderson, Wyman's replacement, and Kaluza's replacement, Don Vidrine—showed up and joined the discussion. The emerging consensus was that the leak through the annular had caused the pressure spike. But Wyman still felt something was wrong.

It was now just past 6 p.m. Randy Ezell had been on duty for twelve hours. The VIP meeting he was expected to attend would begin at seven. But Randy told Jason he'd stick around to help them figure out what to do about the negative test.

"Why don't you go eat?" Jason said.

Randy hesitated. "Well, I can go eat and come back."

"Man, you ain't got to do that," Jason said. "I've got this. Don't worry about it. If I have any problem at all with this test I'll give you a call."

Randy studied Jason closely. He'd known Jason ever since Korea. Jason was not only his top lieutenant; he was one of his closest friends on the rig. Randy felt he could read Jason just by his body language, and now what he saw was confidence. He had no doubt that Jason had what it took to deal with the situation, and would be as good as his word and call at the first sign of trouble.

"Well, maybe I will go eat then," Randy said.

He would never see Jason again.

The night crew decided to conduct a second negative test. They were still debating how do to it when Vidrine pointed out to Kaluza that his shift had ended an hour earlier, and suggested he get out of there and get some rest. You can call Houston on the way out and let them know where we're at, Don told him.

Bob went off. But a few minutes later he was back.

Don did a double take. "Bob, what are you doing here? Go on to bed."

Bob shook his head. "No, he told me to come back and stay with you for the negative test. He wanted both of us up here."

It was a tricky situation. Their problem was that the heavy spacer that had infiltrated the test area was creating pressure that registered on the drill pipe. That pressure would mask any pressure flowing into the well from the hydrocarbon deposit, which defeated the purpose of the negative test.

But they thought they had a way to get around that. Instead of using the drill pipe to conduct the negative test, they could use the kill line. The kill line connected at the bottom of the BOP, beyond the seal on the annular preventer, but still several thousand feet above the end of the drill pipe, where the leaked spacer was wreaking havoc. The pressure created by the weight of the spacer should have no effect on the kill line—which meant if they *did* see pressure on the kill line during the test, it would have to be coming from formation pressure pushing into the well.

They began by releasing the pressure that had built up in the system. Based on their calculations, the compression of the liquid around the drill string should result in 5 barrels of backflow out the top after it was opened. The 5 barrels flowed into the tanks, and then kept flowing. In the end, 15 barrels came out. When they closed the valve, the drill string pressure quickly rose back to 790 pounds per square inch, fell, then slowly rebuilt to 1,400 pounds per square inch over thirty-one minutes.

The results were odd. It wasn't clear, even with the leaked spacer down in the test area, why the pressure reading should rise, fall, then rise again. But they decided to just focus on the kill line.

The kill line was opened. Mud flowed and spurted. The pressure released.

Now they pumped seawater into the kill line to make sure the line was full. Then they opened it again. A small amount of water flowed back out, then the flow stopped. The pressure read zero. They watched for thirty minutes and saw no increase.

"Go call the office," Vidrine told Kaluza. "Tell them we're going to displace the well."

Finally, they were ready to declare the negative test a success.

They were wrong.

A month later, an analysis of well data would suggest that not only did 50 barrels of heavy spacer fluid leak through the annular preventer and fall into the test area, but the rest of the spacer, more than 350 barrels' worth, had not been pumped high enough. During the second negative test, the bottom of the spacer level was invading the top of the test area, including the bottom of the kill line. The reason the kill line pressure fell to zero and stayed there may not have said anything about what was happening down the well. It may have meant nothing more than that the kill line had been plugged with a thick and sticky mix of Form-A-Set and Form-A-Squeeze.

CHAPTER SIXTEEN

SAILOR TAKE WARNING

1900 Hours, April 20, 2010

Macondo Prospect

As the sun dropped toward the rim of the ocean, the horizon caught fire. A handful of rig crew—the poets among the pragmatists—stopped, as always, to admire the reddish orange glow spreading across the western sky. It never got old, reminding them just how close their work brought them to nature's immense power. They were more akin to shepherds than factory workers in that way.

It was the kind of moment Dale Burkeen appreciated. The big bear of a crane operator was just beginning his shift as the horizon launched its nightly light show, and nobody had a better view than Dale, sitting in the cab of his crane, 185 feet above the water. The Horizon was an odd place—the heaviest of heavy machinery, a factory on pontoons, but nonetheless surrounded by an unbroken sweep of ocean wilderness under a dome of the biggest, and often bluest, sky anywhere. He often wished he could share the moment with Rhonda and Timmy. On his last shore leave, Dale had sat on his front stoop, Timmy tucked in the space between his dad's legs looking like a kangaroo pup in a pouch, wishing he could come to the rig with Dad.

Dale had promised a camping trip when he returned home as compensation. Now, as he watched the red smolder in the western sky, he missed his family. This day was his and Rhonda's eighth wedding anniversary, and his thirty-eighth birthday was just four days off. But the celebrations would have to wait.

If Daun Winslow paused to appreciate the sunset, it didn't distract him from mentally outlining the upcoming meeting he had arranged for the VIPs and the Horizon's top managers. It was the main business of this visit, allowing them to get together face-to-face instead of being projected on a video screen. The agenda, meant to be informal, was still rather full. They would review goals for the coming year, talk about maintenance issues and the scheduled drydock renovation of the rig in 2011, reemphasize the need to take a proactive approach in dealing with the danger of dropped objects (maybe they'd even get around to revealing the mystery of what the letters in DROPS stood for). Mention would be made of Randy Ezell's nomination for a company excellence award (for which Randy would no doubt take a lot of affectionate ribbing). And there would be a presentation honoring the Horizon's amazing record of safety—those seven years without a single lost-time incident.

Just before the meeting was to begin, Jimmy Harrell straggled in from the driller's shack where Daun had left him an hour earlier in intense discussions about the negative test.

"Everything all right up on the rig floor there? Get everything sorted out?" Daun asked.

Jimmy gave a thumbs-up.

After the negative test was declared a success, the annular preventer was opened. The pumps cranked up and began again forcing seawater into the hole to flush out mud and spacer that remained in the riser.

The pumping continued for nearly an hour, then slowed as the mud handlers watched for the spacer to appear at the surface. As the pumping rate decreased, the flow from the well into the fluid return tanks should have decreased in step. Instead it increased. The return tanks filled rapidly, and a gauge on the driller's panel registered the increased flow. Amid all the activity, and the security of knowing the well had been sealed, nobody noticed.

Between 8:58 and 9:10, the volume of mud coming out of the well exceeded the volume of the seawater going in by about 2,400 gallons. When the spacer appeared at the surface, the pumping stopped altogether for five minutes, long enough for a "compliance" contractor to test the material and certify it safe to dump overboard, directly into the ocean. The pressure on the drill pipe steadily increased.

Jason called over to the *Bankston* and told the crew to stand by to take on the rest of the mud.

There would soon be nothing but seawater between the rig and the well.

Things were winding up on the drill floor. Chris Pleasant, subsea supervisor, looked at his watch. It was 9:10. "Jason, I'm done," he told the toolpusher. "I need to go work on the BOP crane and get it ready and inspect it to pick up the BOP before—before, you know, we unlatch."

"Okay," Jason said. "You got your tensioners where they need to be?"

"Everything is perfect," Chris said.

In the accommodation block, the VIP meeting was finishing up with the presentation of the Horizon's award for safe operation. When that was over, Horizon department heads went off to bed or to their tour assignments, leaving the four shoreside execs to debate what to do next.

Buddy Trahan was an old hand on Transocean rigs, and there wasn't much he hadn't seen. But Daun thought Pat O'Bryan and David Sims, the BP contingent, might enjoy going up to the bridge to get a sense of the maritime side of things. On a lot of these tours, the VIPs didn't bother to go up there. They toured the rig floor and the drilling machinery, because that's what the rig was to them, a drilling machine.

It struck Daun that they didn't give enough credit "to the individuals who worked the marine systems and whatnot."

Curt had technically finished his shift at six, but the captain is never off duty in the same way as other department heads. He's always on call in an emergency. And of course he had attended the VIP meeting, and now said he'd be happy to escort them. They all tromped up the stairs to the bridge, a large square enclosure directly beneath the helipad on the forward, port side of the rig. It had prominent windows, but especially at night, the view from inside was almost totally dominated by an impressive array of computer screens and instrument panels. Squint and you could imagine being on the bridge of the starship *Enterprise*.

Curt introduced them to the watch officers: Yancy Keplinger, the senior DPO, and his assistant, Andrea Fleytas. While Andrea kept her eye on business, Yancy gave a quick explanation of the various screens and control panels, the most impressive of which looked a little like an ICBM launch panel. That was appropriate enough, since pushing the right buttons could launch the rig's version of the nuclear option—the emergency disconnect system, always referred to simply as EDS. It was a noun, but it was most often used as a verb, as in, "If we ever have to EDS . . ." The EDS was the last resort. The right combination of buttons would send hydraulic fluid charging down a line to the BOP, where it would activate the blind ram shears, the "pinchers" Jimmy had spoken of, driving them through the well shaft, slicing the drill pipe and sealing the well. This was a no-going-back option, as it would disconnect the upper half of the BOP stack from the lower half, and the rig would be freed to move away from the well, dragging five thousand feet of riser and half the BOP with it. Best case, recovering from an EDS would take days, maybe weeks. Worst case, they'd have to abandon the well. In any case, it wasn't an option to exercise lightly.

Yancy moved on to describe the principles of the dynamic positioning system, which kept the rig hovering above an absurdly small patch of ocean.

The VIPs ended up clustered around the DPS simulator in the middle of the bridge. The simulator was essentially a very sophisticated computer game modeling all the forces involved in getting the rig to move, or making it stand still. The simulator had an actual justification for being there—in the many months of standing still in benign conditions, the DPS officers might want to stay in practice for a storm, or the moment when they'd need to move the rig to another well.

The system also allowed them to practice for contingencies. With a few clicks of a mouse, the wind could be kicked up a few knots and a thruster sent out of service. It made a very cool—and challenging—game. They loaded the simulator with seventy-knot winds and thirty-foot seas, and just to make it more interesting, two thrusters down. Then they switched it to manual mode to see if any of them had the chops to maintain the rig on location.

Pat O'Bryan gave it a shot for about half an hour, then they dumbed it down a little and let David Sims have a try. As they stumbled through the exercise, Yancy did his best to give them some instruction, but it wasn't going well. They were wreaking all kinds of virtual havoc on the Horizon.

When the VIP meeting ended, Randy Ezell went down to the galley to get something to drink, then down another deck to his office. He looked at his watch. It was 9:20, fifteen and a half hours after his shift had begun. But he didn't feel his work was finished. He called up to the rig floor and got Jason on the phone.

"Well, how did your negative test go?"

"It went good," Jason said. "We bled it off. We watched it for thirty minutes and we had no flow."

"What about your displacement? How's it going?"

"It's going fine," he said. "It won't be much longer and we ought to have our spacer back."

"Do you need any help from me?"

Jason repeated what he'd said a few hours earlier on the rig floor. "No, man. I've got this. Go to bed."

Randy was tired, for sure, and he would do just as Jason said. But he knew that if it hadn't been for the VIP tour and the dinner meeting, he would still be down there with Jason. It was something he was likely to think about for the rest of his life.

Dave Young had spent most of the day fussing with a dysfunctional thruster, the real-life version of what the VIPs were contending with on the DPS simulator. He'd worked his twelve hours, then attended the VIP meeting, and he wasn't done yet. When the mud in the riser was done being displaced with seawater, they would insert the final cement plug, the top plug, which would finish off Macondo for good, as far as the Horizon was concerned.

They could have set the top plug with the drilling mud still in the well, just after the negative test. The protective layer of the mud would have remained in place until the plug was poured and set, when the plug itself would have become a bulwark against a gas infiltration deep in the well. But BP had opted to remove the mud to the *Bankston* first. Dave would have to wait. One of his duties as chief mate was to be the custodian of cement and mud additives—essentially any heavy, noxious powder that was fluffed and pushed with air from tall steel silos in the columns of the rig. When the cement job started, he'd make sure the system had enough air pressure and all the valves were lined up in the right order to send to the cement unit the powder needed to mix up whatever brew was on the well plan.

Dave tidied up some things in his office on the bridge, then decided to go up to the rig floor and see when they thought they'd be ready for the cement. He was hoping he'd have enough time before the job started to get a little rest. He passed the VIPs hooting it up around the simulator, walked out the starboard door, and climbed up to the rig floor looking for Jason and Dewey.

The first thing he noticed was the pumps. They'd stopped.

"What's up?" he asked.

"It's going to be a little longer than we originally thought," Jason said. "At least a couple hours. We're having some issues with differential pressure we need to check out."

Dave had only a vague idea of what differential pressure issues entailed, but the "couple hours" part of it was plenty clear.

"See you in a couple of hours then," he said.

He went back down to the bridge, put his radio in the desk,

then headed down to get some rest. On the way to his cabin he stopped in the subsea office. It was just after 9:30.

Chris Pleasant was in there, filing some documents on his computer.

"The cement job's going to be a couple of hours yet," Dave said. "They've got some issues with the well."

Allen Seraile, an assistant driller, was in the office with Chris flipping through channels on the TV. Just then the clicker froze in his hand, and he cocked his head, listening. Now Dave heard it, too, the splattering of a sudden downpour on a metal roof. Allen said, "Chris, what's that water?"

Mike Williams was talking with his wife on the phone in his electronic technician's shop when something came on the PA speakers in his office. He wasn't listening, but his wife was. In the background, she heard a natural gas level being announced from Sperry-Sun, the gas monitoring contractors. She asked if he needed to get off the phone to go take care of it.

"Nah," he said. "It's just an indication to make everyone aware of what the gas levels are. We've gotten them so frequently on this well I've become immune. I don't even hear them anymore."

The levels were read off as a number. When the number read 200—parts per million—it was the cutoff for all chipping, welding, grinding. That was when Mike started paying attention, when he knew to stop cutting wiring, doing anything that might make a spark. When that happened he felt more irritation at being delayed than concern about a threat to his safety.

Then he heard a hissing noise and a thump. His first thought was that for some reason they were running the riser skate, which carried riser sections onto the rig floor. The skate ran up right behind his office, and the operators were always slamming it against the mechanical stops in the back, hard enough to shake his office. That could account for the thump, and he assumed the hiss was from the hydraulic leak inevitably caused by banging the skate around like a toy.

"Hey," he said. "I need to go check this out and see what's going on. Make sure we don't have hydraulic oil going everywhere."

He hung up, and almost immediately he heard something beeping through his ventilation system. The vents crossed over between his shop and the ECR—the engine control room—next door. *Beep, beep, beep, beep, beep, beep, beep.* It was continuous, and he knew that it must be the local alarms on the ECR's panel, but he had no idea what kind of alarms. They kept coming, one on top of the other. Now he knew there was nothing routine about this. His heart pumped a rush of blood to his head. What was going on? He was still trying to rationalize it. Did he have a process station acting up? Were these false alarms?

His mind was whirling, trying to put all the pieces together, the thump, the hissing, the beeps. The only conclusion he arrived at was that he needed to get up from his desk and go find out exactly what the hell was happening.

Jason could feel the well boiling now. All thoughts of differential pressures or any other fancy explanations blew away. The gas was coming. It was that simple. It had entered the well and moved up the column. As it rose, the pressure decreased until the liquefied gas became gaseous. At that moment, it began to expand, rapidly, exponentially, pushing everything before it through miles of mud and seawater. Now it was almost there. As he lunged past the assistant driller for the BOP panel he shouted, "Call Randy!"

Randy Ezell had just switched off his overhead light and was beginning to doze with the TV on when the phone rang. He hit his alarm clock light and it said 9:50. He grabbed the phone. It was Steve Curtis, the assistant driller.

"We have a situation," Steve said, his voice strange and tight. "The well is blown out. We have mud going to the crown."

Randy felt the terrible truth of a cliché he'd heard all his life. He couldn't believe what he was hearing. When he tried to speak, he found himself sputtering. "Do y'all have it shut in?" he asked.

"Jason is shutting it in now," Curtis said, and his voice cracked a little. "Randy," he said, "we need your help."

Randy was already out of bed grabbing some pants and trying to stuff his feet in his boots. "Steve," he said. "I'll be—I'll be right there."

Alwin Landry, master of the *Damon Bankston*, was in the bridge of his workboat finishing up some log entries when his DPO said, "There's mud or something coming out from under the rig."

Landry's first thought was that a hose had busted. Then he looked out the starboard window and saw dark liquid shooting through the top of the derrick. It wasn't a spray, it was an explosion. The mud spattered down on his deck, along with some dead birds, knocked out of the sky by the force of the raging geyser of mud. He got the Horizon bridge on the radio. "We're having a well control problem," a tense voice told him. Then another voice came on telling him to go to his five-hundred-meter standby position.

Landry had a problem, too, now. The *Bankston* was just forty feet away from the edge of the rig, and he still had the hose on board taking in mud. The hose was so big and heavy it took a crane to lift it. He was going through options in his head when he looked back at the rig and saw a flash.

CHAPTER SEVENTEEN

"SOMETHING AIN'T RIGHT"

2145 Hours, April 20, 2010

Block 252, Mississippi Canyon, Gulf of Mexico

Down in the engine control room, about the time that Mike Williams put a call through to his wife, Chief Mechanic Doug Brown was seeing to some end-of-shift paperwork. The control room was a steel box with two windows and doors on either side leading to the huge 10,000-horsepower engines, and from there Doug kept an eye on the computer panels that monitored engine temperature, pressure, and rpm's. With him were his boss, First Assistant Engineer Brent Mansfield, and two motormen, Terry Sellers and Willie Stoner, rank-and-file guys who handled the minutiae of keeping the engines roaring, as they were now. The two diesels on either side of the control room, Number 3 and Number 6, were on line and humming through the inch-thick steel fire doors that separated the engines and the control room.

They all heard the hiss. Doug thought it sounded like a large air hose springing a leak.

The four men looked up, around the room, then at each other.

From their expressions it was clear that none of them could identify what the sound was or where it was coming from. Gas alarms began to beep. They half expected to hear an announcement from the bridge over the PA system, but there was nothing. Just beeping and hissing. Finally the radio crackled to life. It was Captain Curt telling the supply boat *Damon Bankston* to detach and move away from the rig.

That broke the spell. Stoner walked over to the radio and turned it up. The hiss had become a roar, like a jet engine. Just as he turned to go back to his chair, engine Number 3, the engine on the port side, began to rev.

Doug said, "Something ain't right."

He turned to the console to study the screen. At the very bottom of the panel, multiple emergency shutdown (ESD) alerts began to flash. Now Doug understood how much danger they were in. The ESDs flashed when multiple gas sensors were triggered in a given area. They were designed to automatically shut all air vents to stop the spread of the gas, which was colorless and odorless, so it could be difficult to know if the dampeners worked. Except . . .

Engine Number 3 revved higher. Methane is also a powerful and explosive fuel. If it had gotten into the engine room . . .

Doug began to run through the awful chain reaction in his mind. Engine 3 was screaming now. Why didn't it shut down? Two automatic engine trips designed to shut down the engines when they went into overspeed should have kicked in by now. He knew they worked because he tested them periodically, forcing the engines into overspeed just to watch them shut down. But the Number 3 wouldn't quit. It kept revving faster and faster, moving through the octaves to the highest pitch Doug had ever heard, then kept going until it was so high he couldn't hear it anymore. He felt he was about to reach some inevitable conclusion when the power surge tripped a breaker and the lights and computers blinked out.

"We're dark," Doug said. Then there was a deafening bang and the heavy steel door to the port engine room blew off its hinges. The door tilted at an odd angle. For an infinitesimal moment, moonlight streamed benignly through the door's porthole. Then everything shifted. The door tore through the control room, slamming Brent Mansfield into the corner, fracturing his skull and tearing away a ten-inch-long slice of scalp. Doug was blasted with the unyielding impact of a freight train. The floor opened beneath him and he dropped into a hole. He was aware that he was in the subfloor, in a tangle of cables and wires. He wondered what was happening. He was confused. He felt pain. Now he heard the Number 6 engine screaming on the starboard side. As he tried to stand, a second explosion slammed him violently back down the hole, and the collapsing ceiling followed him in. He lay there covered in trash and tiles, listening to screams and cries for help, smelling smoke but seeing no fire, wondering when the next explosion would come.

As Mike Williams pushed back from his desk, still thinking about his wife, the computer monitor blew up in his face. All the lights in his shop exploded and fine shards of glass fell like sparkling rain. Now he *knew* they were in trouble. He reached down to grab open the door. Just as his hand touched the handle, the engine on the other side of his wall revved to a level he could have never imagined possible, spinning so fast it just . . . stopped. The awful hissing noise turned to a whoosh and a bang. The door broke its six stainless steel hinges and slammed against his forehead, pushing him clear across the shop. When he came to, he was up against a wall, the door on top of him. His first clear thought was, *This is it. I am going to die right here, right now.*

He didn't die. Within seconds, the room filled with smoke. Mike couldn't see and he couldn't breathe. He managed to push the door off him and crawled across the floor to the opening where the door had been. He reached in his pocket and found a pen flashlight. It worked. He put it in his mouth but still couldn't see anything. Why couldn't he see?

He knew he was in the passageway between his shop and the forward door to the engine control room. He made his way by feel, staying on his hands and knees because he figured if there were any oxygen, it would be at floor level. As he reached the next door, he reached for the handle. It exploded.

Williams flew back thirty-five feet into another wall. His arm didn't seem to be working, and neither did his left leg. He still couldn't breathe or see. The CO_2 from the fire extinguishers was overwhelming him. He knew he needed to get out, to find some air. He crawled back down the hallway and into the engine control room. In his blind progress, he felt himself bumping into a body. He couldn't tell who it was, and there was no movement. He decided that whoever it was was dead, and even if not, he was in no condition to help. He could barely help himself.

But he could crawl, and that's what he did. Halfway across the room, he saw a dim light.

In that same dim light, Doug Brown saw a person crawling across the floor. It was Mike Williams. Somehow, the electronics tech had ended up in the engine control room creeping over the debris and wreckage, screaming that he was hurt and had to get out of there. Mike crawled past Doug, heading toward the back of the control room, where the hatch had blown open and light was streaming in. Now Doug could see that Willie Stoner was crawling toward the same hatch ahead of Mike. Doug fell behind Mike, and all three crawled into the light and the air of the stern lifeboat deck. They were gasping for air when the other motorman, Paul Meinhart, called to them.

"I need some assistance. Brent is injured."

They could hear Brent moaning. Doug wanted to go to Brent, but Mike Williams was in bad shape. Blood was streaming down his face from a wound on his forehead. He kept screaming, "I've got to get out of here!" Doug decided to take him up to the bridge, their muster station, at the other end of the rig while Willie went in to help Paul with Brent.

As Willie came back into the ECR he couldn't see anything at first, but he could hear Paul pulling debris off Brent and tossing it aside. Paul was helping pull off a heavy piece of something—he couldn't even tell what it had been—when Brent lunged up, stumbled, and fell forward, knocking Willie to the floor. Willie pushed himself back up. He and Paul got a hold of Brent and half carried, half dragged him outside.

In the light, Brent saw the blood dripping from his face. "What's this?" he asked, as if he had never seen blood before. Then, as if realizing something, he said, "It hurts."

There was a broken water pipe dribbling water above where the door had been. Paul and Willie used some of the water to wash away the blood on Brent's face as much as they could. Brent looked like he was in shock.

"Calm down," Willie said. "We're going to go ahead and make our way to the bridge." As they carried him up the stairs to the main deck, where the derrick should have been, they instead were met by a swirling tower of flame.

On either side of the rig, just aft of the derrick, the port and starboard crane towers rose sixty feet off the main deck. From his perch in a roomy cab atop the port crane, operator Micah Sandell had seen the mud shooting up and instantly understood he was looking at a blowout. He had shouted into his radio, telling his roustabouts to get to the front of the rig, away from the derrick. But the mud eruption stopped abruptly. Just as he took a deep breath of relief and thought, "Oh, they got it under control," mud spurted out of the degasser, a gooseneck pipe venting out to the deck aft of the derrick. Micah knew then that nothing was under control. The degasser was where mud was routed if it had been saturated by methane down in the well. The toolpusher must have switched the outflow to the degasser when he saw the kick coming, a split-second attempt to handle the kick and avoid polluting the Gulf. Still, it had been the wrong choice. The kick was far too powerful, and the degasser had been instantly overwhelmed. If the incoming hydrocarbons had been diverted through a different pipe that led out over the ocean instead, he would have guaranteed a serious pollution incident, but deadly gassy smoke wouldn't now be belching out of the gooseneck loud enough to rupture an eardrum and making the deck into a time bomb waiting for a spark to ignite it.

That didn't take long. Something beneath Micah exploded, and flames burst out above a shed on the starboard side of the derrick. He jumped up and turned off the air conditioner in his sealed cab, not wanting to suck in any of that gassy smoke, whatever it was. He didn't know if he should get out and run for it or stay.

On the far side of the rig, Dale Burkeen faced the same decision. He forced himself to focus. He knew that in an emergency, an unsecured crane was a catastrophe waiting to happen. He also knew it was his responsibility to direct his crew in firefighting, and he had to get down and suit up in his fire gear. Everything in him told him to flee, but he resisted. First he had to move his crane's 150-foot boom into its cradle in order to lock it down. The seconds ticked by slowly, agonizingly, as smoke and gas billowed higher and higher below him, closing in on his cabin. When the boom was two-thirds of the way down, it became clear the whole process was taking too long. Just like Micah Sandell, crouched in his cab 250 feet away on the other side of the rig, Dale had a decision to make. He couldn't wait any longer. He was a large man, but he dashed as fast as he could to the spiral staircase that led down to the deck and began to pound down the steps.

While Dale ran, Micah remained frozen in indecision in the cab of his crane. A second, bigger burst knocked Micah to the back of the cab. A fireball swirled around him. He fell to the floor, waiting to die. He put his hands over his head and said the simplest and most urgent prayer a man can say: "No, God. No."

Like a miracle, the flames died back from where he was and began shooting straight up the derrick. His paralysis transformed into certainty. He took off out the door of the crane and started down the stairs. When he was nearing the deck, another blast knocked him to the floor. He was almost surprised when he found he could stand, but the surprise turned to a determination to escape. He ran for the lifeboats.

Dale had been about halfway down when the entire back deck seemed to rise up and rush at him. A tremendous shock wave ripped Dale from the staircase and dropped him into the looming cloud of smoke.

Anthony Gervasio, engineer on the *Bankston*, saw the flood lights blink out on the rig. Two or three seconds later, a small explosion lit up aft of the derrick. He frantically reached for an explanation that would lead to some ordinary conclusion. Some welding work? A backfire? Nothing fit. He had no idea what to do. He broke for the engine room door and saw, or felt, another bigger explosion and only then did he know there *was* no ordinary explanation. "The rig just blew up," he shouted. And then he remembered they still had the mud hose on board. He called for help and ran out to disconnect the hose. Together they hit the quick-release coupling, untied it, and dropped it overboard.

After the panicked call from the drill floor, Randy Ezell hung up the phone and lunged for the coveralls hanging on the hook by his door, hopping on one leg and then the other to get them on. He pulled socks on his feet and jammed them in his boots. Then he remembered his hard hat was across the hall in the toolpusher's office. When he stepped out into the hallway, he was dimly aware there were people there, but he barely noticed them. He had only one thought, one purpose: Get to the drill floor fast.

He was reaching for his hard hat in the toolpusher's office when the explosion picked him up and threw him against a bulkhead twenty feet away. The lights went out. The hum of the air handler stopped. Nothing moved. It was silent, deathly calm. He noticed he was covered with debris. He tried to get up but he couldn't move. He tried again, but he was pinned. Whatever was on top of him weighed a ton. He felt a surge of fear and frustration. He thought, "Either you get up or you're going to lay here and die."

Randy tried again, but his leg was stuck. He pulled it with every ounce of strength and desperation he owned and it came free.

He stood up and instantly realized standing had been a mistake. A cloud of smoke seared his eyes and choked him. He knew from his safety training to get down low. There is always cool oxygen on the deck and he was determined to find it even if he had to suck air from the cracks in the floor. He dropped to his hands and knees, but he had lost any sense of direction. He just froze and tried to think, *Which way is it?*

Just then, he felt the faintest puff of air, and decided it had to be coming from the hallway. He crawled toward the air, but the living quarters had been demolished, so he had to move carefully over unidentifiable debris. When he made it to the doorway, he realized that the air he had felt was actually methane. Randy could feel it condensing in drops on the side of his face. Knowing a single spark could ignite the gas at any moment didn't make his painful progress any easier, nor did the dizziness and blurred vision that overtook him as the gas was absorbed by his skin. As he crawled forward, carefully feeling ahead for anything sharp or jagged, his hand hit something solid and yielding. It was a body. He couldn't see who it was in the absolute darkness. Then he saw a light flickering in the hall ahead of him. It bounced and faltered as it grew stronger. It occurred to him it was a flashlight held by someone stumbling down the hall through the debris.

Suddenly the flashlight was in his eyes. Randy saw that the man holding it was the electrical supervisor, Stan Carden. They shined the light on the man down on the floor and saw it was the toolpusher Wyman Wheeler. Wyman had been the one who'd had a bad feeling about the negative test, and now here he was, seriously injured in the wreckage. As they were tending to Wyman, Jimmy Harrell emerged from his room in nothing but coveralls. He looked stunned.

"I was in the shower," Jimmy said. "I bent over, and when I

stood up, the walls of my room were gone." He rubbed his eyes, gritting them furiously. "I think I've got something in my eyes."

Randy looked down and noticed Jimmy didn't have any shoes. "Jimmy," he said, "I've got Wyman down right here."

Jimmy said he had to find some shoes and get to the bridge. He went off down the hall, picking his way through the tangled ruins of the accommodations. Randy and Stan were pulling debris off Wyman when another flashlight appeared. It was Chad Murray, the chief electrician. They sent Chad off to find a stretcher, and kept digging. After a while, Randy thought they might be able to help Wyman to his feet and carry him out. He hadn't forgotten the methane, and he didn't want to stay in there a second longer than he had to. They lifted Wyman up and put his arms around their shoulders. He took a couple of steps and grunted in pain.

"Set me down. Set me down," he said.

They did what he asked. Wyman put his head back and closed his eyes.

"Y'all go on," he said. "Save yourselves."

Randy leaned over Wyman and waited for his eyes to open. When they did, he looked into them. "No, we're not going to leave you. We're not going to leave you in here."

From down the hall they heard a weak, strangled call. "God help me. Somebody please help me." The sound seemed to be coming from the entrance to what had been the maintenance office. All Randy could see was a pair of feet sticking out from a pile of wreckage.

They told Wyman they'd be right back and began to pull debris off the pile above the feet. Almost immediately, they hit something solid and heavy. It was a steel door, and it was lying on top of whoever belonged to the feet. As they bent together and pulled it off, they recognized the man beneath the door as Buddy Trahan, the Transocean VIP. Buddy had left the tour at the bridge early and went down to the accommodations. When he passed by the subsea supervisor's office he saw Chris Pleasant desperately calling the rig floor and getting no response.

"What's going on?" Buddy asked.

"Buddy, we got to go," Chris said, jumping up. They took off running, but got separated. Buddy ended up here, underneath this steel door. The door was sticky with blood and as they heaved to push it aside, they saw why. One of the steel hinges had pierced Buddy's neck, leaving a hole the size of a golf ball in the flesh a half inch from the carotid artery. As Randy bent to comfort him he saw that what at first looked like a black shirt was the bare, blackened skin of his back where his clothes had burned from his belt to his head. A long, deep gash on his left thigh pulsed blood onto the floor. Between pulses, Randy thought he could see bone through the torn cloth and flesh. It was hard to understand why Buddy was conscious.

Clearly, he was their most critical casualty. So when Chad came back with the stretcher, they loaded Buddy first. It took all three of them to get the stretcher out into the open because one of them had to go ahead and clear the debris hanging from the ceiling and jutting up from the floor. The space they had lived in for years was unrecognizable to them. When they emerged on deck, Stan and Chad took the stretcher and headed to the forward lifeboats. Randy stayed with Wyman. He'd said he would not leave him, and he couldn't do much, but he intended to at least honor that promise.

CHAPTER EIGHTEEN

MAYDAY

2149 Hours, April 20, 2010

Block 252, Mississippi Canyon, Gulf of Mexico

Andrea Fleytas had been monitoring the dynamic positioning system on the bridge of the Horizon when she felt a jolt. Before she could make sense of it—a rig-shaking shock that came out of nowhere—magenta warnings began flashing on her screen. Magenta meant the most dangerous level of combustible gas intrusion.

Andrea was a triple minority on Deepwater Horizon. She was a woman, a Latina, and she was a Left Coaster. It was hard to say which one made her stand out the most. She grew up in an East Los Angeles barrio, excelled in high school, and graduated from the California Maritime Academy. When she earned her four-year degree at the top of her class, she got this job with Transocean. Not many years ago, the offshore business was an insider's game—you had to know someone in Transocean to have a prayer of getting hired. And that's if you were a man. The oil business was an old boys' club, and everyone knew it. A woman taking that on had to have special self-confidence, and even at the young age of twenty-three, Andrea had it.

But when faced with the multiple magenta alarms, she froze. Her training had taught her that when more than one warning light flashed, she needed to hit the general alarm immediately. Now there were so many lights flashing, she didn't know how to respond.

Andrea's supervisor, Yancy Keplinger, felt the jolt, too. He abruptly left Captain Curt and the puzzled BP VIPs he'd been tutoring on the dynamic positioning simulator in the middle of the bridge and strode to where Andrea sat at the main control console a few steps away. He punched a few buttons to point a closed-circuit television camera to the starboard side of the rig. On the monitor, he saw mud streaming out of a diverter pipe.

The phone rang and Andrea answered it.

It was the rig floor. A voice said they were fighting a kick from the well, then hung up. The magenta alarms kept flashing, indicating one part of the rig after another.

Then all the screens blipped off. The room went dark. For a moment, there was an almost complete silence.

A backup battery kicked on and the screens came back to life. Andrea and Yancy checked the monitors. No engines, no thrusters, and except for the emergency battery on the bridge, no power whatsoever. A flash of light lit the walls followed instantly by a bang so loud it was impossible to tell if it was heard or felt. Yancy called the rig floor. Nobody answered. Now Andrea jumped up and hit the general alarm. She grabbed the radio and began calling over an open channel, yelling, "Mayday, mayday. This is Deepwater Horizon."

Curt heard Andrea repeating the mayday. It was a direct violation of chain of command. Regulations permitted only the captain to give the order to call mayday.

He came up behind her: "I didn't give you authority to do that."

The first blast shook chief engineer Steve Bertone out of bed. Bertone was thirty-nine. His black hair was shaved close on the sides and flat on top, and he had the shadow of a goatee on his chin. He'd been on the Horizon since 2003, and he'd been chief engineer for seventeen months. He was a direct and earnest man with an intelligent intensity about him. He wasn't excitable, a "sky-is-falling" type. But that huge thump that kicked him awake must mean that *something* was falling. He thought that the wire holding the fifty-ton top drive block must have parted, and the shattering crash was the block coming down the derrick. His first thought was that he needed to get to the bridge. He pulled on clothes, work boots, grabbed his hard hat and life vest, then ran out into the hallway. Four or five people were clustered at the bottom of the central stairwell, frozen, looking up as if they couldn't understand what they were seeing. Steve followed their gaze. There was so much debris, he couldn't tell if the stairs were there or not. Either way, it was completely impassable.

Steve had to break the spell. "Take the port forward or starboard forward spiral staircases!" he hollered. "Go to your emergency stations!"

Steve took the port staircase up to the bridge and went to his station, the rear port-side computer. He saw immediately what they were dealing with. No power and no thrusters. He picked up the phone and punched 2268, the engine control room, before he realized there wasn't a dial tone. He put the receiver down, then picked it up. Still no dial tone. He did it again. There were no phones.

"We have no coms!" he shouted. He ran over to the starboard

window and looked back at the derrick. In shock and denial, he hadn't even registered the second explosion. He was still expecting to see a jumble of steel and pipe that had crashed down on the rig floor. Instead, where the derrick had been, he saw a great wall of fire.

Doug Brown and Mike Williams stumbled from the collapsed engine control room onto the rear of the rig. When they got their eyes cleaned out enough to see, they noticed several things. Where there had been a lifeboat deck, there was no lifeboat deck. Where there had been two lifeboats, there were no lifeboats. Where there had been a walkway and handrails, there was only a ragged ledge dropping off into darkness. If they had stumbled one more step, they would have been in the ocean. Mike was still horribly confused, but he knew this: Something really, really bad had happened, and it wasn't going to get any better any time soon.

Now Doug was sure Mike was in shock. Blood was still pouring from the wound in his forehead. Doug decided that whatever else he had to do, he first had to get Mike to a medic. Now that the rear of the rig had blown off, they had no choice but to make their way clear across the rig to the bow.

They took the back steps to the main pipe deck. As they cleared the last step, they were almost knocked over by a blast of heat followed by an unbearable hissing sound, like something emitted from the lungs of a gigantic and predatory creature on the verge of roaring. Flames shot more than 200 feet into the sky, consuming the derrick from legs to peak. They understood now for the first time what had happened. Doug felt sick as he looked at the place where the drill floor should have been and saw only fire. He wanted desperately to imagine a way the drill crew could have gotten out in time, but he couldn't do it.

His skin began to smolder in the intense heat from the fire. They would have to walk right past the derrick to get to the bridge, and the lifeboats, if those amenities still existed. They turned to the left, upwind of the flames, and made their way forward, shielding their faces from the heat with their hands.

Steve Bertone ran back to his computer thinking that the engine should be restarting. The engines were designed so that in a power disruption, after twenty-five to thirty seconds the two auxiliary engines would come on line and automatically start feeding power to the thrusters. But there was no indication that had happened.

While he was puzzling over the negative indicators on his screen, Steve heard the door to the bridge bang open. He turned to see a man covered in blood from head to foot leaning toward Curt. The man cried, "We have no propulsion, we have no power, we have no *ECR*!"

Curt looked uncomprehending.

"You need to understand," the blood-covered man said, pleading. "We have no ECR. It's gone. It has blown up. Engine Number 3 for sure has blown up. We need to abandon ship *now*." Steve suddenly recognized the voice. It was his number two, Mike Williams.

"You need to calm down," Curt said. "Just calm down, sit down, we're working on it."

Doug Brown had come in the door behind Mike Williams. "He needs a medic," Doug said. Steve ran over to them and saw that the source of all the blood was a deep laceration across Mike's forehead. Steve hollered, "Where are the medical supplies?" Someone said they were in the restroom, in the back of the bridge. Steve couldn't find any gauze there, so he grabbed a roll of toilet paper, sprinted back, and put the roll to Mike's head. "Hold this there," he said.

He went back to his station still expecting to see the engines coming back on line—he just couldn't believe those engines were gone. But there was still no power. The realization took hold. They were on a dead ship above a blown-out well and there was precious little they could do about it.

Daun Winslow stood between the two forward lifeboats looking up at the derrick, which was blazing like a 250-foot high Bunsen burner. He was wishing he'd see more people coming out of the accommodations.

The decks were covered with mud and a substance with the consistency of snot that had to be the remnants of that double dose of spacer. It was slippery as hell, and hell was exactly what it looked like all around him. People were screaming that they had to get off the rig. People were crying, begging God not to let them die. As Daun tried to help guide them forward on the treacherous surface through the mass confusion, Curt appeared above him, just outside the starboard door of the bridge. Curt caught Daun's eye and waved his arm in a gesture Daun took to mean that he wanted him to get everyone back inside the accommodation block, as if that would be the safest place for them to be. Clearly Curt hadn't yet realized the extent of the damage there. Daun ran up the stairs to tell him what was going on.

"Captain Curt," he said, "the accommodations are destroyed. We've got to get people to the boats."

"Okay," Curt said. He may have been about to say more but Daun heard someone screaming, threatening to jump overboard.

Daun raced back down the stairs. A man Daun had never seen before was clinging to the outside of the handrail.

"Hey," Daun said, "where are you going? There's a perfectly good boat here."

The man leaned away, toward the ocean.

Daun came closer and made eye contact. "Do you trust me?" he asked.

"I don't know you," the man said.

Mark Hay, the senior subsea supervisor, appeared beside him. Daun pointed to Hay. "Do you trust *him*?"

"I don't know," the man said. "I'm not sure." But he seemed to lean back in, away from the seventy-five-foot drop into the ocean. Gently, Daun and Mark Hay took the man's arm and coaxed him back onto the rig.

"Go on," Daun said. "Get on the lifeboat with everyone else."

Then he turned and looked back toward the derrick. It was still burning as high as ever, and that didn't make sense. The well should have been sealed by now and the BOP separated into two parts and disconnected from the well, cutting off the gas that fueled the flames. What was going on? Why hadn't they pushed the disconnect button? He ran back up the stairs to the bridge.

After Chris Pleasant told Buddy Trahan to run, they both ran, but in opposite directions. Buddy headed back into the accommodations, and Chris toward the moon pool to try to find out what was happening. When he got to the exit he ran into Chad Murray, the electrician. "Man, I wouldn't go that way," Chad said. "Something bad just happened."

"What?" Chris asked, but Chad had already run off, back down the hall.

Chris turned up the stairs taking two at a time to get to the main deck and the drill floor. Then he saw the fire. He turned around and headed straight for the bridge, and his BOP control panel. As subsea supervisor on duty, he was the person responsible for executing the EDS, which is what he damn sure intended to do right now. Curt stepped in front of him.

"I'm EDS'ing," Chris told him.

"Calm down," Curt said. "We're not hitting the EDS."

Chris stepped around Curt to the panel. Don Vidrine, the company man, was standing there looking at the indicator lights. "We got the well shut in," he said.

Chris looked and saw the lights on the panel that indicated the lower annular was closed, but clearly it hadn't done the job. The rubber seal on the annular preventer had not been powerful enough to stand up against a full-roar blowout. And the gas must have already been past the BOP, into the riser, by the time Jason had managed to close it. The fuel was still flowing to the fire. Chris knew he had to go to the last resort, the shear ram, the most powerful intervention on the BOP—hydraulically powered pistons that would drive through the drill pipe, detach the BOP, and seal the well, cutting off the continuing flow of fuel to the fire and freeing the rig to move away.

"Don, I have to get off," Chris said.

"Get off," Don said. "Push the button."

Chris pressed down the enable button, held it, then hit the fire button.

The indicator lights flickered from green to red, open to closed, just as they were supposed to do. Chris felt a wave of relief, until he noticed another set of indicators that measured the flow of hydraulic fluid that drove the shear ram through the drill pipe.

The gauges read zero. The hydraulics were dead.

The bridge door slammed open and Daun Winslow came in. Over his shoulder, Chris heard the captain ask Daun for permission to EDS. As Transocean performance manager, Daun was the highest-ranking company employee on the rig.

"You haven't already?" Daun asked. "Yeah, hit the button."

Someone on the bridge yelled out, "He cannot EDS without the OIM's approval."

As if on cue, Jimmy Harrell burst through the door onto the bridge.

"EDS, EDS!" Jimmy was shouting as he ran toward them.

Chris looked up from the panel.

"I already did," he said.

Steve Bertone ran over to Chris. "I need confirmation that we've EDS'd."

"Yes," Chris said.

"Chris, I have to be certain. Have we EDS'd?"

Chris said yes again, and pointed at the lights in the panel.

Steve didn't wait for part two, the part about the hydraulics not flowing. He turned and hollered across the room.

"Captain, do I have permission to start the standby generator?"

"Will it give me fire pumps or any propulsion?"

"No," Steve said. "It's going to give us lighting, and it will give us the ability to bring engines back on line later."

His assumption was that now that the ram shear had sealed the well, the fuel still in the riser would soon burn out, then they could use the standby generator to run the air compressors needed to restart the surviving main engines.

"Yes, go," Curt said.

The standby generator was located near the derrick, very near the fire. Steve turned to go. Dave Young came over with a pair of handheld radios so Steve could call for help if the fire trapped him. They turned them on but couldn't get them to work, even standing five feet apart. They verified they were on the same channel, and tried again. Still nothing.

"Don't worry about it," Steve said, laying the radio down on the table.

As he opened the door to leave, Mike Williams pushed the door closed.

"You're not going alone, chief."

"Don't be crazy, you're bleeding."

"You can't do this alone. If I'm not going, you're not going."

Steve shrugged.

"Well, come on, then," he said.

Paul Meinhart, the motorman, grabbed Mike's shirt. "I'm coming, too," he said. And they went in a line like that, the three of them, back to the fire. Steve looked up where the top of the derrick should have been and saw nothing but flames. He kept nearly losing his footing on the slippery deck. To get there they had to pass by the BOP storage area. He looked into the moon pool. It was filled with flames.

But the standby generator room was pitch black. Mike put a penlight in his mouth so they could read the startup procedure posted on the wall. Steve flipped the switch from automatic to manual, and hit the reset button and the start button.

Nothing happened.

He tried it again, the reset button then the start. Again, nothing happened.

"I have battery power," Mike said. "We have twenty-four volts."

Paul was standing over by the watertight door, looking out at the flames.

"Shut that door," Steve said.

He had to think. If they had twenty-four volts, why wasn't the engine turning over? He flipped the breaker shut, then reopened it. He ran back to the panel and again tried the reset and the start. Absolutely nothing.

"That's it," Steve said. "Let's go back to the bridge. It's not going to crank."

They opened the door to start back. It was like walking into an oven.

Dave Young was torn between staying on the bridge and going back on deck. The bridge was still in chaos, Curt seemed overwhelmed, but as chief mate his responsibility was to direct the emergency response and firefighting. He had to go.

As he was leaving, Andrea stopped him. She told him the captain had told her to stop the mayday call.

"Do it anyway," Dave said, running out the door.

When Dave got to the emergency gear locker, the muster point for the emergency team, only one person was there, a roustabout named Christopher Choy. Apparently, everyone else on the fire team had ignored their training and already gone to the lifeboats. The fire was spreading, and there were small explosions crackling all over the rig. Oddly, Dave didn't feel afraid, just desperate at the thought that there might be people trapped by the flames, and that any reasonable chance of rescue was quickly vanishing.

Dave knew he had to hurry.

As soon as he began to put on his fire suit, the padded, fireproof pants, the helmet, mask, and air tank, he decided that would take too long. He just grabbed the coat and ran toward the column of flame. Chad Murray, chief electrician, came running the other way.

"Dale Burkeen is down," Murray said. He said he figured Dale had been trying to climb off the crane when the blast knocked him to the deck. He wasn't moving.

Dave started to sprint toward the crane. Another blast knocked him off his feet and drove him back twenty feet. He picked himself up, stunned that all his body parts seemed present and operational. He could still run, he discovered, as he dashed toward Dale's prone figure. Dale was on his back, unconscious, bleeding. Dave felt for a pulse. He couldn't find one, but knew he couldn't stay there trying. He tried to pick him up, but Dale was much larger than even an average rig worker, and Dave couldn't budge him. He ran back to the gear locker, but there was still nobody there but Chris Choy, who weighed all of 150 pounds. But Chris was young, twenty-three, and strong—he'd been on his high school power-lifting team back in his hometown of Tyler, Texas. And Dave knew they were running out of time.

"Dale's down on the deck, and I can't move him by myself," Dave told him.

They retraced Dave's steps to the spot on the drill floor where Dale had fallen. Small explosions burst all around them. The rig was beginning to yield to the heat. They heard things popping and falling, crashing and banging. They knew that if you cooked a derrick long enough, at some point it would come down. They just hoped it wouldn't come down now, on top of them. They kept moving. They must have been less than fifty feet away when a jet of flame flared in front of them. A blast of heat knocked them back, blocking their path. Without fire hoses and water, without a team of men, without proximity fire suits, there was nothing they could do.

"Chris, we can't get to him," Dave said. "The boats are going to launch. You need to go."

Chris looked like he was about to protest, then turned toward the boats.

Dave looked back into the fire once more, hoping he'd see another way around. There was none. He kept thinking that he'd been right *there*, he'd had Dale in his grasp, and he'd had to let go. Now it was too late.

He sprinted back up to the stairs. As he barreled through the door onto the now dim and smoky bridge, the captain asked him why he wasn't fighting the fire. Dave didn't think Curt was getting it. He grabbed hold of the captain's shirt and led him outside. They could see the growing column of flame from the bridge windows, but they could not feel the heat. Dave waved at the flame and screamed, "Look! We have no pumps, no thrusters, no way to get off the well. There's nothing we can do. We have to abandon ship."

CHAPTER NINETEEN

ABANDON SHIP

2157 Hours, April 20, 2010

Block 252, Mississippi Canyon, Gulf of Mexico

The forward lifeboat deck was bright as day in the light of the blazing derrick. As the stunned, dazed, and injured gathered at their assigned muster points, they only found singed and twisted holes in the rig. They kept stumbling forward to the only two lifeboats remaining.

The boats were shaped like booties, fully enclosed, capsule-like cylinders with benches on both sides long enough to fit 73 110-pound Koreans hip to hip and knee to knee—but they could only fit from 45 to 60 of the much larger Americans, depending on the number of 250-pounders aboard. Any way you looked at it, with two of four boats gone, the math didn't quite work now. Sixty times two was still some short of the 126-member crew.

When Doug Brown boarded, an assistant driller he'd known since the Horizon left Korea was checking names off a muster list. The man looked at Doug blankly.

"Name?" he asked.

People were screaming, "Why can't we leave now" and "I don't want to die."

Doug tried to remain calm, but he was scared. The fire was growing; the rig was coming apart.

Not more than fifteen minutes earlier, Micah Sandell had been sitting in the cab of the port-side crane. Now, waiting to board the lifeboat, that seemed a lifetime ago. Micah was still shaken from his narrow escape, the screaming and hollering of people who wanted the boat to leave without him weren't helping hold back his fear. A man screamed through a megaphone, but what he was saying was lost in the roar bursting from the derrick and the lungs of the terrified. Some were trying to count heads and load the wounded, but others were yelling, "Drop the boat, drop the boat!"

Micah attempted to follow the procedure learned in every-Sunday drills, but some were broaching the line and jumping into the boats. Others had frozen, hypnotized by the flames, unable to move or respond.

Gregory Meche, a mud engineer, was astounded by the size of the fireball roaring into the sky, high as a skyscraper. He wasn't panicking, exactly. He thought the muster situation was fairly controlled given the circumstances. It looked just like a fire drill, only with everybody involved at once, and a little more chaotic. But as he stood there, Greg felt the passing seconds weigh on him. The thought of sitting in one of those crowded, closed boats not knowing when or if it would leave made him queasy. He hadn't been waiting there for more than five minutes, but he just couldn't stay another second. He had to move.

And he did. Down the stairs to the deck beneath—the smoking deck—and over the edge into blackness.

Something flashed in the mercury vapor lights. The crew on the *Bankston* leaned as far over the rail as they could for a better look.

Was it a life ring? Or something else? Then an arm came out of the water. It was a man trying to swim toward them. Anthony Gervasio broke for the rescue boat on the back deck. He saw a second jumper out of the corner of his eye as he ran. Cook Kenneth Bounds had seen the jumpers, too, and sounded the man-overboard alarm, alerting Captain Landry on the bridge. Landry ordered Mate Jeffrey Malcolm to launch the rescue boat. The deck was slippery so Malcolm didn't flat-out sprint, but he got there as fast as he could without taking too big a chance. Gervasio got there first, and by the time the third and fourth bodies hurtled from the rig, he was already trying to get the rescue boat lowered and ready to go. He had to concentrate on freeing it from the belly straps that held it in place. He pulled on them to get them to release, consciously keeping his motions deliberate, ticking down the familiar process as he went. Remove the charger from the battery. Make sure the painter line was clear. Tilt the motor. Lift up the seat and switch on the battery. Lower the boat. Lower the motor. Start the engines and make sure they were running right. He knew they would be, because he'd just used the boat in a drill, but he didn't want to skip a step. He didn't want to make any mistakes. This was no drill.

The fireball over the derrick flared larger. The sudden surge of energy transmitted to the people waiting at the lifeboats made them push forward, scrambling over each other for a seat. Roustabout Stephen Stone climbed into Lifeboat Number 2, strapped himself in, and waited. Time dragged like fingernails on a chalkboard. Every passing second made him squirm in excruciating discomfort. Stephen knew he was going to die. It was odd how he could just sit there and wait, wondering how it would happen. Maybe the derrick would topple on them. Or maybe they would all suffocate from the smoke in the lifeboat.

Some people who were already on the boat must have been thinking the same thing. They jumped up and ran back off onto the deck. Some kept going down the stairs and over the rail. Meanwhile, someone was still futilely trying to get a head count. People kept screaming, "Drop the boat, drop the boat!"

A tall, lean man with thinning hair leaned in from outside.

"Don't launch yet," he said. "You have to wait."

It was Daun Winslow. He put one foot into the boat and kept another on the deck, as if to dare them to launch.

Daun coaxed more people into the boat. Two men walked up carrying a stretcher. The man's injuries were so terrible that it wasn't until Daun was helping load it that he recognized the man on the stretcher as Buddy Trahan, the Transocean colleague he'd flown out to the rig with just hours earlier, hours that now seemed to belong to some alternate universe. The lifeboats were designed for people who could sit on benches, not people laid out on stretchers. Buddy was loaded across a row of laps. He'd been in and out of consciousness, but he stirred into a fog of awareness. He could hear the people around him gasping at his injuries—"look at his leg, look at the hole in his neck." The thought that he might die soon made him think of his three children. The nine-year-old would soon find it hard to remember him, and the teenagers would never get over losing him at such a vulnerable time. He felt himself sinking again, but before he blacked out, he saw the tower of flame. It shocked him into a clear thought, a stab of pain. Nobody on the drill floor could have survived that. They were all gone.

"Get in and shut the door and let's go," someone inside the boat said. There was still a crowd of people waiting to board. Daun watched the frightened faces flickering in the fire's light.

"We've got plenty of time," he said.

The words were barely out of his mouth when a drilling block and traveler, 150,000 pounds of equipment, fell fifty feet from the derrick, crashing into the deck only yards away.

Daun backed out of Lifeboat Number 2.

"Take it down," he told the coxswain.

Dave hurried to the lifeboats. When he got there, Number 2, the port boat, had already launched. Daun Winslow was in the door of Number 1 and it looked like they were preparing to follow. Dave was making calculations in his head. There were about a half-dozen people still on the bridge, and he knew that Chad Murray and Steve Bertone were bringing Wyman up from the accommodations.

"Hold the boat," Dave said. "We've got a man down, and it may take us a while."

"Don't go far," Daun said.

But as Dave climbed back to the bridge he saw Daun duck down into the lifeboat, the door closing behind him. Dave knew he would have done the same.

As Steve, Mike, and Paul made their way back to the bridge from the auxiliary generator room, Mike looked over the starboard side of the rig and saw Lifeboat Number 1 motoring away. They picked up their pace.

When they got to the bridge Curt said, "I've given the order to abandon ship. We can't fight this fire. It's time to leave."

Yancy and Andrea were still at the radios coordinating the mayday response.

"That's it. Abandon ship!" Dave hollered at them. "Let's go, now."

As they funneled out of the bridge Dave was realizing he was the only one without a life jacket when a hand tapped him on the shoulder. It was Andrea. "I saved this for you," she said.

Now, as they all took off running down the steps to the forward lifeboats, they saw Lifeboat Number 2 descending.

Randy Ezell waited for another stretcher to come. Time had warped into an unrecognizable form, an enormous and implacable void that had neither shape nor movement. Wyman Wheeler lay beside him, quiet for the moment. Randy was sure Wyman's leg was broken and when they'd tried to lift him he'd screamed in pain and said his shoulder hurt. He listened to Wyman breathing and waited for footsteps. He had no idea how long it had been—five minutes? a month?—before Stan Carden and Chad Murray returned.

They lifted Wyman on the stretcher and moved as gently as they could, but he grimaced in pain as they maneuvered through the obstacles. When they were outside, they walked past the flames to the bow. As they came up to the lifeboat deck, they saw the captain and the rest of the bridge crew running down the steps.

The lifeboats are gone, the captain said. *We're taking the life rafts.*

The rafts were essentially twenty-foot-wide kiddie pools with a tentlike canopy, roughly the size and shape of an early Mercury space capsule. They were stored in plastic cases that could be cranked up on a davit, rotated over the side until clear of the rig, then automatically inflated with a nitrogen cartridge. To board, you had to step out over the side of the rig into the raft through a flap in the tent cover. That was the theory.

The theory didn't account for the fire burning out of control above and below the raft launching point. A twenty-foot bulkhead behind the lifeboat station provided their only shield from the gargantuan conflagration, and it was quickly losing its effectiveness. The rig was coming apart. Projectiles were sizzling past in all directions. They discovered as they came to the edge of the deck that a back draft was coming up the underside of the rig and converging right on the lip where the raft would dangle as they tried to board it.

Explosions had become nearly continuous now, like the crescendo to a fireworks display. Every second's delay seemed potentially fatal. Dave took charge of deploying the raft, but as he frantically cranked the davit handle lifting the raft, it yanked to a halt. They had missed a rope tethering the raft to the davit. Someone yelled for a knife, but nobody had one. A no-knife policy was one of Transocean's safety measures. Mike Williams grabbed on to the rope and followed it to where it was attached with a small metal shackle. He tried to unscrew the shackle by hand, but it was too tight.

The heat of the fire was so intense now they could smell hair singeing and feel exposed skin burn. Mike reached into his pocket and pulled out one of his electronic tech tools. It looked like an oversize pair of nail clippers. Using them like pliers, he managed to unscrew the shackle and free the raft.

When they finally got the raft over the side, it jolted down and stopped three to four feet below the edge of the rig, tilting slightly and spinning away. It wouldn't be easy to get everyone loaded, and the stretcher was going to be dicey. It might tip the raft. Or worse, they might drop it into the ocean. . .

Dave realized that someone needed to be inside the raft to control the swinging and pull it close to the side of the ship. He climbed atop the handrails, waited for the raft's door to turn toward him, and leapt inside. The raft now dangled by a cable. The smoke billowing down the side of the rig had a clear path into the raft's door. Dave stood at the door and leaned his body across the gap between the raft and the side of the rig to pull the raft in closer.

Steve looked at Wyman on the stretcher, and then at Curt. He thought he heard Curt say, "Leave him!" Paul Meinhart, standing beside Curt, didn't hear anything like that. But Steve wasn't going to leave Wyman no matter what. He ran to the stretcher.

"Let's get him to the life raft," he said.

Chad Murray jumped into the raft to help Dave receive it. Dave struggled to hold on to the rig's edge. He didn't just smell burning rubber; he could taste it. He realized that the heat radiating from the fire looming above them had been met by an updraft from beneath the rig. The steel deck had been shielding them from it, but both currents of superheated air now converged precisely where the raft dangled over the edge. He felt his flesh cooking, but he willed himself to hold on.

Steve fell to his knees on the deck and fed the end of the stretcher to Dave and Chad. His knees burned through his pants and he could feel the heat through the leather work gloves he didn't even remember putting on. The raft lurched as Dave and Chad hauled in the stretcher. Wyman started to scream, "My leg, my leg!" When the stretcher disappeared into the raft, Steve jumped in next, then Andrea and the others, leaving Curt at the edge of the rig, Yancy Keplinger and Mike Williams still standing behind him. The raft was melting; they all knew it.

Then two explosions occurred, one right after the other, blasting the air from their lungs. Andrea, who had been the calm, strong voice of the Horizon projecting the maydays across the Gulf, started to cry and scream, "We're going to die!" Steve had to agree with her. So did Dave and Chad. They were going to cook right there.

The raft filled with black smoke. Everyone ducked to the floor to find breathable air, everyone except Dave. Clutching the rail, all he could do was take small sips of foul air and wait for it to be over.

"Take it down," Curt told Dave. Dave reached up to release the brake. The raft rocked forward, then plunged fifty feet. As it fell, it caught on the painter rope designed to tether it to the side of the rig, flipping first to one side, then the other. They tumbled around like clothes in a dryer. Andrea fell right through the door into the water. The raft plopped down beside her and, sputtering, she grabbed on to the edge. The smell of burning rubber and petroleum was overwhelming. A hundred feet away, directly beneath the rig, the water was on fire.

Curt, Mike Williams, and Yancy Keplinger, alone now on the burning Horizon, watched the raft plunge into the ocean.

Yancy couldn't believe the captain had sent the raft down without them. "What about us?" he screamed.

Curt had thought he would take one last look around to make sure everyone else had gotten off the rig, but the flames had grown to the point where there was nowhere to look.

"I don't know about you," Curt said, "but I'm going to jump."

Mike eyed the two remaining life rafts. He thought about how long it had taken to deploy the first one with ten people to help. Now it was just the three of them, and Mike felt his strength failing fast. They'd never make it.

"We can stay here and die," he said, "or we can jump."

He remembered his training: Reach your hand around your life jacket, grab your ear, take one step over the edge, look straight ahead, cross your legs, and fall. The problem was that now there was a life raft down there. They couldn't just step and fall without crashing into the raft.

"We're going to have to run and jump," he said.

Curt took three steps and leapt.

"He just did it. You've got to do it," Mike said.

Yancy did.

Now Mike was alone. He backed up a few steps, closed his eyes, spoke to God about his wife and daughter, then took off running.

At seven minutes past ten, Paul Johnson, the shore-based rig manager for the Deepwater Horizon, got a call in Houston from Paul King, a Transocean colleague.

"You know, I don't want to alarm you," King said, "but we're getting mayday calls from the Horizon."

It can't be, Johnson thought. My phone is right here. Someone would have called.

Just then, his phone began to beep.

When the raft splashed down, Dave found himself in the water. Steve Bertone jumped out the exit door to join him. The raft started drifting underneath the rig into the spreading pools of oil floating on the surface, and toward the furnace raging at the center. They paddled around the side, grabbed the rope, and swam away from the burning water, away from the rig. Chad Murray and Paul Meinhart splashed in behind them. All four kicked and stroked and pulled on the rope. As Steve side-stroked away, he looked up at the rig above him, clouded in smoke like a fog. A pair of boots and work pants came shooting through the smoke. It was Curt. He plunged in five feet away. A second pair of boots came flying out of the smoke and hit the water about ten feet away. It was Yancy Keplinger.

Steve could tell they were making progress. His angle on the rig was widening. He'd gotten to a point where he could see above the edge of the rig deck just in time to watch a man sprint full speed and leap off the edge. He was still running in midair as he began to drop. Just before he splashed into the water, he seemed to look directly at Steve. It was Mike Williams.

Curt and Yancy swam to the raft to help pull and shove it away from the fire. Suddenly, they couldn't move it any farther. Someone in the raft yelled, "Oh, my God, the painter line is tied to the rig!" Steve looked over his shoulder and saw a taut white line stretching from the raft back toward the rig until it vanished in the smoke. Chad Murray started screaming for help. Dave and Steve followed Chad's eyes. Fifty or sixty yards away, he saw the rescue boat, and two flashing lights in the water. The boat converged on the lights, and as Steve watched, the flashing lights became people in life vests, as first one, then the other, was hauled aboard.

Then the boat turned and raced toward them.

Mike fell for what seemed like a very long time. He went deep into the ocean, silent and black. He looked where he thought up might be and saw an orange glow far above him. Then he began to rise. He burst to the surface and his lungs filled with air. His heart pounded with relief. He'd made it. He was going to live.

And then he noticed that his skin was burning. His face and eyes were burning. He was burning all over. Was he on fire? The water felt like grease and smelled like diesel fuel. And then he saw the fire on the water, not far from where he was and coming closer. He had to get out of there. He swam. He kicked and pulled and kicked and pulled and something strange happened. The pain in his head went away. He couldn't feel anything. He was wondering if he might be dead when he heard a faint sound. It was a voice, saying, "Over here, over here."

He was still trying to make sense of it when his life jacket seemed to rise of its own power, carrying him with it. A pair of strong hands hauled him up and flipped him into the open bow of a speedboat. He still didn't know where he was or why, but he didn't care.

The fire crept closer to the raft, no more than twenty-five feet away now, as the rescue boat approached. The people splashing in the water and sitting in the raft were all screaming for a knife. The boat cut the engine to neutral and glided toward them. A man came to the bow holding a large folding knife. Curt swam to the boat, took the knife, then swam back to the raft and cut the rope.

Now they all started swimming the raft toward the rescue boat. They threw the man standing at the bow a line. He caught it and wrapped it around a cleat. The boat shifted into reverse, backing them away from the fire, away from the rig.

With his last strength, Dave swam to the side of the boat. The final few feet seemed like miles. His arms and legs felt like dead weights, heavy enough to pull him to the bottom. He went limp in his life jacket and waited. Hands reached down and pulled him in. All the adrenaline his brain could produce had been used up. The fear he'd kept one step ahead of all night finally overtook him. He felt the pain in his body for the first time. The burns on his arms and head throbbed. He looked back at the Horizon burning in the night, and he thought:

"Holy shit! Holy *shit*!"

The Deepwater Horizon had blown up, and he was alive.

CHAPTER TWENTY

MUSTERING

2200 Hours, April 20, 2010

Block 252, Mississippi Canyon, Gulf of Mexico

When all the survivors were aboard the *Damon Bankston*, hurting and bleeding and disbelieving, they watched the flames a half mile distant and wondered who hadn't made it off.

The *Bankston* crew had pulled seventeen out of the water, including Andrea and the group in the tangled life raft. The rest had come over in the two lifeboats, but in the chaos, nobody was really sure who had been on the boats and who hadn't, or even how many there were. Dave had nothing left inside him; he was done. But he knew he had one more task to perform. He had to get people who only wanted to collapse into oblivion to line up and give a definitive roll call. They had to know.

On the *Bankston*'s main deck, the boat's crew had created a makeshift field hospital where the most severely injured were being tended by medics. For the other survivors, *Bankston* crew were giving up berths, and in many cases the clothes off their backs, to provide whatever small measure of comfort they could. Dave found some reserve he'd had no clue existed in him and managed to keep upright and moving among his hollow-eyed rig mates, trying to make sure they had water or blankets or something warm and dry to wear.

As he went, he ran through a list in his mind of those he'd seen near the drill shack on his last visit before the well blew; he searched the shadowed faces huddled on the deck looking to find them there. Some appeared just when he'd given up hope. Others never did. There was one face he saw over and over again, but he knew for sure he would not find it on the *Bankston*: the big, open, smiling moon face of Dale Burkeen.

They took roll, twice. A quick-thinking radio operator on the Horizon had brought the rig's muster list, so they had that to work with. On the first attempt, they came up with fifteen missing, but by the second time through, there were just eleven names unaccounted for. When Dave considered the names, and that each of them had been on duty on the rig floor when the well blew, he knew that all the search boats in the world wouldn't make any difference.

After the muster, Steve Bertone circled the *Bankston* to check on his engineering crew. He found them all, except for one: Brent Mansfield. He proceeded to the only place he hadn't yet looked, the makeshift hospital. He found Brent lying on a stretcher on the floor, though it took a while to realize that that was him. Brent's tangle with an angry steel door in the engine control room hadn't gone well. His head was entirely covered with bandages and gauze. He had an oxygen mask over his mouth and nose and a neck brace.

Brent was crammed into a tight space near a bunk holding another injured man. Steve slid in between the two, determined to keep Brent awake. Steve had always heard that if people who'd suffered head injuries slept, they could fall into a coma or die. The truth was a bit more complicated than that, but Steve wasn't taking any chances. Every time Brent started to drift off, Steve nudged him awake by adjusting his oxygen mask and talking about whatever came to mind.

As badly as Brent was hurt, the man on the bunk above him looked to be in even worse shape. It was Buddy Trahan. Buddy kept drifting off, too, so Steve doubled his efforts and tried to keep both men conscious. It was like spinning plates on sticks. Every time he'd focus on one, the other would start to wobble.

Some time after the Coast Guard helicopters arrived around 11:30, a rescue swimmer appeared. He was all got up in a Navy SEAL type of wet suit. They'd clearly expected to be fishing people out of the water, but it hadn't worked out that way. Now he was going to begin the evacuation of the seriously injured.

"Who's the critical?" he asked.

The medics pointed to Buddy. They brought in a gurney and Steve moved around to the backside of the bed to help get Buddy moved. He put one hand on Buddy's hip and one on his shoulder and began to roll him as slowly and gently as he could onto the gurney. At the first tilt, Buddy bellowed in agony about his leg. The blanket fell away and Steve could see that Buddy had a deep gash on his left thigh, below which his calf was mangled and oddly twisted. Buddy's fingernails were gone, and the hole in the side of his neck was just the most terrifying of a net of lacerations all over his body.

As Steve braced himself to roll Buddy a little farther toward the gurney, the medic standing on the opposite side of the bunk gasped. Steve looked where the medic's horrified gaze had fastened and saw that Buddy's back was burned black from belt to head.

When the medics took Buddy out to the helicopter, Steve stayed until they came for Brent, then he made the rounds of his crew again. After that, he climbed up to the *Bankston*'s uppermost level, where he sat alone and looked out across that narrow yet infinitely vast stretch of water between where he was now and where he had been.

The Horizon refugees were crowded together in whatever makeshift space could be found, the ship's lounge, its galley, the open deck, and in whatever berths the *Bankston* crew had given up to them. There were still some who didn't have dry clothes, and many were barefoot. They were stuck here until the Coast Guard released the *Bankston* from search-and-rescue duties, which wouldn't happen until long after almost everyone on board had lost hope that there was anyone to search for or rescue.

Randy Ezell looked around bitterly. There must have been twenty-five boats on the scene, plying the waters in search patterns. He could see no reason for them to have to just sit there, helplessly. Every minute within sight of the raging holocaust on the Horizon was torture to him, disfiguring him in a way he wasn't sure he'd ever completely recover from. Yet it went on and on as they sat there through the night, denied comfort and rest, forced to linger in the light of the very flames that had consumed their *brothers*. He'd never felt that word resonate more truly.

Randy knew all the names left unchecked on the muster, of course. As senior toolpusher, he had held himself responsible for all of them, even the mud engineers, who had worked for a contractor. Three of the missing were his assistant drillers, Donald Clark, Stephen Curtis, and Wyatt Kemp. Donald had a wife, two sons, and two daughters. Today was Stephen's fortieth birthday, a fact that wouldn't make things any easier for his wife and two teenage kids.

Wyatt was just twenty-seven, and only recently promoted. He had a pretty young wife, Courtney, and two daughters, Kaylee, three, and Maddison, just four months old. Kaylee and her mom had a ritual of counting down the days until Wyatt's return on a calendar. The mud engineers, Gordon Jones and Blair Manuel, both worked for the contractor M-I SWACO, but were as nearly family as anyone else. Gordon's pregnant wife, Michelle, was due any day now. At fifty-six, Blair was as irrepressible as a teenager about the summer wedding he and his fiancée were planning.

Karl Kleppinger, Jr., was thirty-eight, a roustabout and a veteran of Desert Storm with a teenage son named Aaron, who had special needs, and a wife named Tracy, who took care of them both. Floor hands Shane Roshto and Adam Weise were just getting started in the business, and in life. Shane was only twenty-two but already shouldering big responsibilities. He was putting his wife, Natalie, through college and they had a three-year-old boy, Blaine. Adam was just twenty-four, a high school football star who'd gone straight from high school to the rig.

Then there was Dewey Revette, Randy's forty-eight-year-old driller, bright and good-natured as the day was long, and on a rig, it was longer than most. And Dale Burkeen, the big-hearted crane operator whom everyone loved.

As for Jason Anderson, the irony that April 20 was to have been his final full day on the Horizon was too painful to consider. Just hours from now, when the sun rose, Jason was supposed to have been gathering his bags and preparing to helicopter off to his new assignment, senior toolpusher on another rig. There would be no helicopter ride to a bright future now. Randy had looked forward to staying in touch, trading e-mails from one senior toolpusher to another. Instead, now—and he didn't doubt for the rest of his life—Randy would replay Jason's last moments in his mind. As Jason fought to control an uncontrollable well, his final request had been for Randy to come help.

Not fifty feet away, Doug Brown stared out at the fire burning above the black plane of the ocean and thought of the day, many years earlier, when he had first laid eyes on the Deepwater Horizon.

CHAPTER TWENTY-ONE

GOING HOME

April 21, 2010

Gulf of Mexico

Sometime after midnight, Curt found Dave on the main deck of the *Bankston*. Dave was in a kind of trance state now, operating in a part of his brain that probably never slept. Curt had gone from the rescue boat straight to the *Bankston*'s bridge, where he'd showered and changed into borrowed clothes. Captain Landry had asked Curt to coordinate the effort to fight the fire, but insisted he clean up first—the powerful kerosene scent wafting off his clothes was distracting the bridge crew. Now Curt urged Dave to do the same.

No way, Dave said. None of the others had been able to shower, and as long as they couldn't, he wouldn't, either.

Suit yourself, Curt said, and he went back up to the bridge.

It was early morning when the Transocean rig manager for the Deepwater Horizon, Paul Johnson, reached his OIM, Jimmy Harrell, on the *Bankston*'s satellite phone. Paul was Jimmy's direct supervisor and had always thought highly of him. He asked Jimmy how he was, but Jimmy could barely speak. Paul thought he might be crying, but he wasn't sure.

Jimmy said he was still having trouble with his eyes. Insulation from the destruction of the accommodations had gotten in there and it felt like it had never gotten out. He was struggling to see and his hearing was off.

Paul tried to reassure him but found that difficult. Then he asked the question that had been burning a hole in his brain all night.

"Jimmy, what happened out there?"

"I don't know, Paul," he said. "She just blew. I don't know what happened. She just blew."

Now Paul was almost certain Jimmy was crying.

"Don't worry about it, Jimmy," he said. "We'll find that out later on. Just take care of yourself."

And then he hung up.

The *Bankston* was finally released from search-and-rescue duty and got under way at 8:13 the next morning. As they left the rig in their wake, it was still burning as fiercely as ever and was beginning to tilt to one side. But even now the survivors weren't taken directly to shore. They were about to begin a zigzag voyage across the Gulf without receiving any explanation of where they were going or when they would finally reach their destination.

Throughout the night, they had been left to struggle with thoughts of wives and children and family desperate to learn their fate, but whom they were unable to contact. The voyage stretched on interminably. First they stopped off at another drilling rig, Ocean Endeavor, where *Bankston* crew took on medical supplies and cigarettes, which were seen as one of the most urgent needs.

They also dropped off Daun Winslow, who had been up all night charting post-blowout logistics with Transocean colleagues in Houston. Daun transferred to another workboat and immediately headed back out to the Horizon to direct an attempt to activate the failed BOP shear ram using ROVs, and to supervise firefighting efforts until contractors arrived to take over.

Both endeavors were doomed to fail miserably. When the hydraulic and communications cables from the rig to the BOP pods—which ran from a spool near the moon pool—were destroyed in the fire, the BOP should have gone into automatic "dead man" function. The loss of signal should have triggered bottles of pressurized gas to drive shut the rams and close the well. Later analysis indicated that low charge in a battery in the blue control pod and a faulty solenoid valve in the yellow pod rendered the dead man function inactive. Twenty attempts by Daun and his team to directly force various blowout preventer rams shut using the ROV all failed to stop the flow.

Fireboats poured hundreds of thousands of gallons of water and flame-retardant foam on the Horizon, possibly contributing to its top-heavy instability. The rig would capsize and sink at 10:22 a.m. on April 22. It would pick up speed as it descended and make the mile-long plunge to the bottom within minutes. Long before it crashed into the mud, the riser ruptured and began spewing crude oil into the Gulf at a rate of as much as two and a half million gallons a day. One plan after another to kill the well or cap it failed. It wasn't until July 15, eighty-six days and an estimated 185 million gallons of spilled oil later, that an effective cap finally halted the flow. Macondo was not declared permanently dead until September 19, after the completion of two relief wells.

The months-long ecological catastrophe—a catastrophe whose consequences will continue to manifest for years to come—would soon overshadow the loss of eleven Horizon crew members and the suffering of all who had been aboard in the final hours of April 20.

For now, as April 21 dawned, their ordeal was not over. The *Bankston* still had another stop to make. Coast Guard and MMS personnel were waiting on the Matterhorn production platform to board the *Bankston*. They had targeted key Horizon crew members for questioning about the sequence of events leading to the blowout and evacuation. The interviews proceeded while the *Bankston* finally headed to Port Fourchon, southwest of New Orleans. When they arrived at 1:30 a.m. on April 22, nearly twenty-eight hours after the blowout, the *Bankston*'s decks remained covered with drilling mud, and some of the Horizon survivors were still barefoot.

Micah Sandell was among the many who hadn't slept. He was beat, desperate to call home, desperate just to be away from there and begin to put this nightmare behind him. But as he walked off the *Bankston* he noticed some Coast Guard officers sitting at a table in front of a row of portable toilets. Before he left, he and everyone else would have to provide a urine sample. Sandell was furious, or as furious as he could be in his extreme mental fog. He stood in one line to fill out the testing forms, and then another to use the porta-potty to fill his little plastic cup. They thought that somehow the blowout might have been his fault?

Everyone knew how ludicrous that was. But Coast Guard regulations required that after any maritime accident resulting in death or more than one hundred thousand dollars in damage, those involved in the incident needed to undergo drug testing within thirty-two hours. The decision was made that rather than try to determine who had been "involved" in the blowout, it would be simpler just to have *everyone* tested.

Simpler for the Coast Guard, maybe.

Doug Brown had ridden up, up, up in the basket as the thudding chop of the helicopter grew into an all-encompassing reality. Then he had been inside. He'd turned his head to find beside him Buddy Trahan. Buddy had been conscious, but fading in and out. Every now and then Doug had heard him moan.

The chopper set them down on BP's Na Kika production rig, not far from the Horizon. He'd been taken to a room set up for triage, then put on another helicopter and sent to the University of Southern Alabama Hospital, where he discovered they'd also taken Paul Meinhart. Doctors ran tests and treated Doug's leg injuries, then released him to two Transocean consultants, who drove Doug and Paul to the Crowne Plaza Hotel in New Orleans.

Getting wheeled into the lobby of a luxury hotel hours after escaping a blowout and a burning rig was a jarring experience. Given how exhausted and shaken he was, it seemed more hallucination than reality. Doug was handed some clean clothes and then wheeled into a debriefing on the blowout with Coast Guard officers. Somehow he managed to participate. He kept telling himself it would only be minutes before he could close a hotel room door on the world and fall into bed.

But when the Coast Guard was done with him, he was taken to another room. There were two men in suits who introduced themselves as legal representatives for Transocean. A court reporter sat beside them with an air of alert anticipation and a transcription machine.

They had a few questions for him.

At 7 a.m. on Wednesday April 21, a Houston attorney named Steve Gordon got a call on his 800 number. The woman identified herself as Tracy Kleppinger. She said her husband, Karl, was on the

Deepwater Horizon, an oil rig in the Gulf of Mexico. Steve knew where this was going. He had been blipped awake at 4:30 that morning by an alert on his smartphone to a CNN newsflash about the explosion on the rig.

Steve specialized in maritime law, and he had represented the families of rig workers before in personal injury lawsuits. He'd recently represented the widow and one-year-old child of a man on another Transocean rig. The man had been working a double shift. He was using hand signals to direct a pipe handler, telling him where to drop a one-ton pipe. It was a hot day without any breeze, so the man found a shady spot to stand in. The pipe handler didn't see the man in the shade. The pipe swung out of control, crushing the man against a steel stanchion. Steve was there to help the widow hold Transocean accountable.

But in this case, Steve's assumption was off.

"I don't want you to sue anyone," Tracy Kleppinger said. "I just want you to find my husband."

When the phone rang at 5 a.m., Alyssa Young was in the shower getting ready to start her day, always a complicated proposition when you were alone in a house with a six-year-old, a five-year-old, and a four-month-old baby. Getting dressed, she saw the phone blinking one message. For a fraction of an instant, she felt a shock, a sudden vivid glimpse of a nightmare she'd managed to forget, that black cloud on the drive to the airport with Dave. But it instantly faded, just like a bad dream can vanish.

Then the day started and the kids decided to have a bad morning. Last night, Dave had messaged her that he wouldn't be able to call, he was busy with some cement job or something. He'd probably decided to call early and catch her before the day started, is all. He'd call back later. She couldn't think about it now. Between getting everyone a different breakfast and getting them out the door, she didn't even have time to turn on the news.

Tracy Kleppinger had gotten a call before dawn. Someone with a dull mechanical voice, almost robotic, said, "I'm calling to tell you that there has been an explosion on your husband's rig. We don't have details at this time. You'll be contacted when we know more." Nobody had called back. She'd found Steve Gordon's name in an advertisement on gCaptain, a networking website for mariners. and called for help.

Steve got a number for the Transocean human relations chief, who took Steve's number and Karl's name and promised to call when he knew something. Steve didn't wait to hear. He kept calling all morning, but learned nothing.

Steve planned for the worst. He chartered a flight for his investigator to go sit with Tracy Kleppinger in her home in Natchez, Mississippi. He'd arrived at her house at noon.

Around 2 p.m., Steve got the news he'd been dreading: "We have now been able to conclude that there are eleven men missing, and that Karl was one of them."

Steve hung up, swallowed, then called his investigator's cell phone to warn him what was about to happen, and admonish him to stay close to Mrs. Kleppinger. Then he called the house.

When he told her that he had bad news, that Karl was one of the missing, he expected her to drop the phone or weep or scream. He didn't expect what happened.

"He's not missing," she said, buoyantly. "They found him!"

She was watching MSNBC. "They're reporting right now that they found a capsule with the missing eleven."

"Oh my God," Steve said. He patched in the Transocean HR chief, and repeated what Tracy had seen on TV: "They have found Karl and the others."

"No," the HR chief said.

"Yes, they have!" Tracy said. "Go to MSNBC.com, you'll see."

"No," he said. "I'm sorry, but we have followed that story, and that is not correct."

Tracy Kleppinger, whose voice had been so alive, so full of hope and relief, just disappeared.

Steve's law partner flew into Natchez and spent the next three days with her. There was still a nominal search operation going on, but nobody really believed anyone would be found.

Within hours of the official cancellation of the search and rescue, on Thursday, April 22, Steve Gordon filed a wrongful death suit in federal court on behalf of Tracy and Aaron Kleppinger.

The funeral was on Sunday, May 2, the day before what would have been Tracy and Karl's eighteenth wedding anniversary.

Shortly after the funeral, Transocean CEO Steve Newman asked to meet with all of the decedents' families. Steve arranged a meeting between Newman and Tracy and her parents in Natchez. Newman apologized to them for what had happened, then he cried, and they all cried together.

When they formally canceled the search and rescue mission at 5 p.m. on April 22, Dale Burkeen's family still couldn't accept that he was gone. His sister, Janet, couldn't stop thinking that Dale would have found some way out. His favorite song was "A Country Boy Can Survive," by Hank Williams, Jr., and his favorite TV show was *Man vs. Wild*, featuring a former British special forces operative demonstrating survival techniques in some of the world's most forbidding locales. Dale was always watching that show.

Janet remembered when she'd tease him about it, and say, "Bubba, when ever can you put that into action?" He'd say, "Sis, you never know what's going to happen to me on the rig. Don't ever give up on me. I'm a survivor. I'll find me something to hold on to and paddle to an island and survive."

So Janet kept thinking the searchers had to have missed him somehow. She just knew he was going to turn up on some little island in the Gulf, sitting under a palm-thatched hut eating bugs and coconuts to survive.

On the grounds of the small-town elementary school Dale attended, there is a small monument to a first-grader who died in an accident on the school playground. A child fell off the sliding board and hit his head. Six-year-old Dale Burkeen was coming up right behind that boy. One of the teachers got rattled and in the confusion of the moment thought it was Dale who'd hit his head. The school called Dale's mother, Mary, and told her that Dale had fallen and "was unresponsive." She rushed up to the school to discover that it was not her son headed for the morgue after all. But the incident sent a chill through the family. It somehow marked Dale as vulnerable, at risk of premature death.

"It just wasn't Dale's time," was how they always told the story in the years to come, as Dale grew into a big, strong country boy who knew how to survive. "It wasn't Dale's time."

Somehow, and God only knew how, they were going to have to learn to accept that April 20, 2010, was.

Alyssa Young was dropping her second child off at school at 9 a.m. when her phone rang.

It was Dave's mother.

"Have you talked to Dave?" she asked. "His rig blew up."

"Are you sure it was his rig?" Alyssa asked, trying to hold down her panic. It had to be a mistake. It had to be.

Alyssa drove to the oral surgery office where her mother-in-law worked. In streaming video on her computer screen, the Deepwater Horizon burned and burned and burned. Dave's mother was pacing and listening to Alyssa make one phone call after another. In spite of everything, she called the rig. It rang busy. She called the Transocean number flashing on the screen. They told her they were still waiting for an update from the Coast Guard.

She looked back up at the computer screen and that unquenchable torch that had been Dave's rig. How could Dave have survived that?

She got a ride home—there was no chance she could drive—and got online. She made serial calls to hospitals in three states, cried on and off, then called more hospitals. By 11:30 she had lost all patience with Transocean and began calling the hotline every ten minutes. How could the news stories be saying there were eleven missing, and Transocean not know which eleven? And if they knew, how could they not tell her? Family and friends filled her house and took the kids outside to play. Alyssa kept searching the Internet for . . . anything. Dave's brother had to finally tell her to just turn off the computer.

At 1 p.m., the ordeal ended. Transocean called to say they had good news: Dave was coming in to Port Fourchon.

She didn't quite believe it until he called at nine that night. He could barely speak; he just kept repeating, "The fucking rig blew up!"

The following night, when he walked through the airport gate in a greasy jumpsuit, hard hat in hand, he had nothing at all except the hundred-dollar bill Transocean had handed him.

"You look like an escaped convict," she told him.

But she knew that it was both of them who had escaped.

After he got home, Alyssa began to follow all the hearings into the cause of the blowout. She had to know what had happened, and she had to believe that it could be prevented in the future if she was ever going to let him return to the Gulf.

Dave was on full-pay leave. He played with the kids and worked on his speedboats to get them in racing prime. When he healed, he raced.

A few minutes after Dave had left for one of his races, Alyssa felt that same sick feeling she'd felt before he left for the Deepwater Horizon on April 14. She called his cell.

He was fine, he said. The race was just about to start.

Twenty minutes later, her phone rang.

"I'm in the ambulance," he said, laughing.

His little boat had flown out of the water at high speed, then flipped. The propeller on the 25-horsepower engine had cut through the Kevlar wet suit and sliced his leg. If it hadn't been for the Kevlar, the leg would have been at the bottom of the Sound.

She called him every name she could think of. She was so angry, she immediately got online and listed all his boats for sale.

He laughed at that, too. But Alyssa was deadly serious. She didn't want him to race anymore. Time passed and using all his charm, he persuaded her to come watch the next one. Dave professed a reformation: he'd drive safely.

"He held back a little," Alyssa said. "But it's just like with the rig. I think it's going to happen again in ten years. That's what humans do, right? They get cautious for a while, and then they forget."

EPILOGUE

A Hole at the Bottom of the World

On April 22, the Deepwater Horizon sank, rupturing its riser, which began a massive oil spill into the Gulf of Mexico. The U.S. Coast Guard and BP estimated the volume of the leak at 42,000 gallons of crude oil a day—roughly enough to fill an Olympic size swimming pool. That accounting turned out to be absurdly optimistic. In the days following the disaster, after multiple failed attempts to trigger the ram shears on the blowout preventer, it began to dawn on an anxious Gulf Coast that an uncontrolled flow of oil might continue for weeks, if not months. Before the end of the next week, with people all over the world watching a live feed, via the Internet, of the billowing leak, estimates of the disaster's size increased fivefold, to as much as 210,000 gallons per day.

President Obama promised to assign "every single available resource" to contain the spreading oil, and vowed to hold BP accountable for the cleanup. For his part, BP chief executive Tony

Hayward launched an aggressive PR campaign with an initial TV ad buy estimated to cost at least $50 million.

"The Gulf spill is a tragedy that never should have happened," Hayward said in his upper-crust British accent as the spot began. The camera moved in for a tight close-up of his sober face and worried brow. BP would take "full responsibility for cleaning up the spill. We will honor all legitimate claims, and our cleanup efforts will not come at any cost to taxpayers. To those affected and your families, I am deeply sorry."

The ad closed with images of the Gulf Coast as Hayward intoned, "We will get this done. We will make this right."

It wasn't going to be easy.

BP pursued multiple strategies for stopping the unchecked flow into the Gulf. The best hope for a quick fix was a 98-ton steel containment dome designed to cover the ruptured wellhead and siphon the oil into a waiting drill ship. The dome was lowered over Macondo on May 7, but soon clogged with slushy a mix of frozen gas hydrates after an ROV collided with the cap and accidentally shut off a deicing system.

Meanwhile, the BP-leased Transocean rigs Development Driller II and Development Driller III began to sink relief wells parallel to the existing Macondo shaft. Once they penetrated beyond 15,000 feet, into the pay zone, the rigs would then drill horizontally and tap into the well at the source of the blowout, where they could pump in enough heavy mud and cement to bury it for good.

But the drilling would take several months, at least, and the unnerving possibility that the gusher would continue unchecked for that long ratcheted up the tension just as executives from BP, Transocean, and Halliburton appeared before congressional hearings in Washington. By that point, estimates for how much oil was spewing each day had climbed still higher, to 800,000 gallons.

Despite the "full responsibility" pledge in BP's ad campaign, it became clear that the company's strategy in the hearing was to blame Transocean for the malfunctioning BOP; Halliburton for the cement failure; and the Horizon crew for misreading the negative test and failing to spot the upwelling oil and gas in time to do anything about it. Transocean and Halliburton joined in the finger-pointing, focusing on BP's suspect well design. Obama called the hearing a "ridiculous spectacle."

Meanwhile, the growing oil slick, now visible from space, was spreading in a lopsided loop around the blowout site. The leading edge of the slick made landfall on uninhabited and environmentally sensitive barrier islands off the Louisiana coast. Soon oil would sully beaches in Alabama and western Florida. Unexplained deaths of marine wildlife began to escalate to many times the normal rate. Though only a small percentage of the carcasses were visibly oil-fouled, the high death rate was almost certainly related to the environmental degradation brought on by the flood of oil blooming in the Gulf.

On May 26, BP embarked on what they called a "top kill," the company's last real hope for ending the gusher sooner rather than later. Even the name of the maneuver had a kind of gunslinging swagger to it, as if John Wayne might come riding in from off screen to solve everything with one well-aimed bullet to the head.

A top kill, though, isn't so much about precision as it is overwhelming force. BP was ready to throw the kitchen sink at the gushing oil, hoping to drown it in a high-pressure barrage of drilling mud and cement—50,000 barrels' worth—pumped from barges on the surface. The technique had never been attempted a mile deep in the ocean, and after more than two days of trying, BP officials had to acknowledge that it had all been an expensive waste of mud. The force of the oil flowing out of the well simply blew away everything in its path.

On Saturday, May 29, BP announced the failure. When the stock markets opened the following week, BP shares plunged 17 percent, wiping out $23 billion of stock equity in a matter of hours. The day of the blowout, BP stock had been trading at $60 a share. By the end of June, the share price would fall to just above $26 a share, cutting the company's value by more than 50 percent in just two months. Transocean's stock price plunged by a nearly identical percentage, exactly 50 percent, from $92 a share at close of business on April 20 to $46 a share on June 30.

If that wasn't a big enough dose of bad news for the two companies, U.S. attorney general Eric Holder announced within days of the top-kill failure that the Justice Department would conduct criminal and civil investigations into the rig explosion and the oil spill. The president wanted "to know whose ass to kick" over the disaster.

Two weeks later, the CEOs of Exxon Mobil, Chevron, ConocoPhillips, and the U.S. arm of Royal Dutch Shell took turns speaking before Congress. They all agreed on one thing: Their companies would never have designed a deepwater well the way BP had designed Macondo. This damning testimony was followed days later by a statement from Anadarko Petroleum, the minority partner in Macondo, that BP had "operated unsafely and failed to monitor and react to several critical warning signs during the drilling of the Macondo well. BP's behavior and actions likely represent gross negligence or willful misconduct."

BP's Hayward and chairman Carl-Henric Svanberg met with White House officials and announced that the company, which had suspended dividend payments to shareholders, would set up a $20 billion fund for damage claims from the spill, and pay $100 million to workers put out of work by a federal moratorium on deep-sea drilling.

When the crew of the *Damon B. Bankston*, the Tidewater workboat responsible for saving so many lives on April 20, had returned to the Gulf in May for their next hitch, it was clear the world had changed. Every oil service port on the Gulf Coast was now full of boats idled by the moratorium. Some of the workboats and rigs in the Gulf would make occasional errand runs for the production platforms still in operation, but for the most part, crew members' time was filled with maintenance chores, busywork. It was an unsettling coda to the heroics of April, though their spirits would be lifted six months later, when, their captain, Alwin Landry, representing all of them, was awarded the maritime industry's top honor: Shipmaster of the Year.

The 98-ton cap was reinstalled in early June, but the 500,000 gallons of oil siphoned away each day seemed relatively insignificant compared to the amount still visibly spewing into the Gulf on the live feed.

By July, tar balls from Macondo began to wash up on the Texas coast, the only Gulf Coast state to have—until then—avoided the spill.

As the oil spread, BP quietly prepared another experimental piece of equipment they called a "capping stack," which engineers hoped might shut off the flow completely. The capping stack was a three-ram BOP, custom-designed to attach to the inoperative BOP left behind when the Horizon sank. The stack had been in the works since shortly after the blowout, but, following the earlier failures, BP managers didn't want to raise too many hopes for what was essentially an experiment. After weeks of construction, the capping stack underwent more weeks of testing on the surface with equipment that mimicked what was still attached to the Macondo wellhead.

Some experts feared that shutting off the well at the top could force out the well walls farther down the shaft, allowing oil to escape into the formation and flow into the Gulf from dispersed fractures in the seafloor, a flow that, once begun, would be virtually impossible to stop.

On July 12, with a grave awareness that this new attempt could do more damage than good, the capping stack was installed on Macondo in a maneuver even more complex and painstakingly executed than the Horizon's original latch-up to the well. Over the next three days, all openings and valves within the device were closed one by one under the watchful gaze of retired U.S. Coast Guard rear admiral Thad Allen, the federal government's national incident commander. With each closure, the pressure inside the new device was carefully monitored.

BP engineers had projected that if the well were compromised, permitting oil to leak into the surrounding formation, the pressure within would never rise above 6,000 pounds per square inch, even with the new device completely shut. If the operation proved a success and the well were kept intact, with the oil flow captured by the stack, the readings would—in theory—rise to between 8,000 and 9,000 pounds per square inch and hold steady. In other words, they would know if the ruptured well had been truly contained only by the digital readout on a computer monitor.

The pressure followed neither predicted course. It rose to 6,700 pounds per square inch before stalling. Anxious watch was kept on pressure readings over the next few days. Seismic surveys of the area around Macondo were analyzed for evidence of seafloor breaches. Ultimately, officials concluded that the well had not been breached, and the lower-than-expected pressure might be explained by the extensive depletion of the oil reserve in the Macondo deposit. The new capping stack had finally stopped the flow after eighty-four days, in which an estimated 4.9 million barrels of oil—more than 205 million gallons—had issued from the blown-out well. Once the well was finally capped, on July 15, officials determined that the actual rate of oil leakage, per day, had been 5,000 times the initial estimate of 42,000 gallons a day.

Three weeks later, on August 4, government scientists reported surprisingly positive news: 75 percent of the spilled oil had been captured, evaporated, burned, skimmed, or dispersed by chemicals and surprisingly efficient natural processing by petroleum-eating bacteria.

But other, independent scientists disputed both the methodology and conclusions of the government study. One dissenter called the claims "ludicrous." On August 16, University of Georgia scientists reported that they had analyzed the federal estimates and concluded that 80 percent of the oil said to be gone from the Gulf was in fact still there. The report cautioned that it was "a misinterpretation of data to claim that oil that is dissolved is actually gone."

In any case, even if the government estimate was taken at face value, it meant that 50 to 60 million gallons of crude oil remained in the Gulf, with undetermined consequences.

The following day the controversy over the government report took a backseat to news that BP had completed a "static kill" by pumping cement through the capping stack to seal the well, though failure-weary officials stressed that Macondo could not be declared dead until the relief wells reached its base and sealed it from the bottom.

The relief well broke through on September 16. The next day, cement was pumped into the bottom of Macondo for seven hours straight. Two days later, the well from hell was finally issued a death certificate.

Meanwhile, evidence had accumulated that the rosy government report on the quick disappearance of most of the oil had seriously understated the continuing threat to the Gulf ecosystem. Controversy erupted over BP's admitted use of 1.8 million gallons of an oil dispersant called Corexit—which critics called both ineffective and toxic to the environment in its own right, more so than other dispersants, but which BP officials said had been chosen for the job because it was available in the huge quantities required.

University of South Florida researchers concluded that much of the oil that had been considered "dispersed" had actually fallen to the seafloor in "a blizzard" of oil particles. The University of Georgia research team announced it had discovered a thick layer of oily sediment covering several hundred square miles of Gulf sea bottom. In some places, the oil was more than 2 inches thick. In October, other researchers reported a plume of oil more than 20 miles long at a depth of about 3,500 feet. Perhaps as disturbing as the existence of the plume was the fact that over several months of observation, it had barely degraded at all.

By this time, BP had unveiled an extensive and highly technical report on its internal investigation into the disaster. Just as BP executives had testified before Congress, the report focused on the misread negative-pressure test, the malfunctioning of the blowout preventer, the faulty cement seal job, and the failure of the rig crew to recognize that gas was surging up the well. Once again, the missteps were attributed primarily to employees of Transocean or Halliburton. The report brushed aside the significance of the BP engineering decision not to use more centralizers in the final cement job, concluding that the fatal inflow of hydrocarbons occurred at the bottom of the well, not along the sides, where the lack of additional centralizers might have played a role in weakening the cement seal. Officials from both Transocean and Halliburton called the BP report unconvincing and self-serving.

Buddy Trahan, the Transocean VIP critically wounded when the exploding well threw a steel door on top of him, arrived at a more personal conclusion. He was suing BP and Halliburton for his injuries, even as Transocean continued to pay his salary and medical expenses, which had exceeded $1.5 million by late summer, around the time that he could finally walk on his own to feet without the aid of a walker.

"The more I learn about this well, the madder I get," he told a reporter for Bloomberg news service in August. The Transocean crew and equipment were not at fault, he concluded. "It's pretty clear to me now it was a screwed-up plan."

Janet Woodson, the sister of Horizon crane operator Dale Burkeen, who died after falling to the deck of the burning rig, had none of Buddy Trahan's intimate knowledge of offshore drilling. But she nonetheless felt rage at the accusations against the drill crew.

Janet heard about the heavily appended 193-page BP report through the local news in Philadelphia, Mississippi. The nuances of the conclusions were all but lost on her. Statements from the report that parceled out blame for the disaster among "multiple companies and work teams" meant, to her, that the drill floor crew, who had spent their final breaths doing whatever they could to react to the emergency, was being saddled with ultimate responsibility.

In late December, as the disastrous year 2010 was finally at an end, almost four months after the BP report was issued, Janet was still struggling with the idea. Just thinking about it made her teary.

"How can somebody sit there and blame the victims when they're not here to defend themselves?" she implored. "That's heartless. I know my brother wasn't to blame. He was on the crane, trying his best to come down and come home."

Like many family members of Deepwater Horizon victims, Janet has perceived in the aftermath of the blowout a kind of abandonment, as public attention almost immediately shifted from the lives lost on the rig to the spreading cloud of oil and its impact on the environment. "Sometimes I feel like we're nothing more than a fish in the ocean to some people," she said.

Every day since April 20 has been a challenge for Janet, but Christmas, the first without Dale, had been harder than most.

"When Dale went off on the rig," she said, "we may not have had Christmas on Christmas Day, but we always had it as a family, whether a week early or late. He always felt bad we had to work it around his schedule, but I always told him, 'It wouldn't be Christmas without you.'"

In 2010, much of Christmas Day was spent visiting Dale's grave with her mother and father. "Last year at this time we never dreamed this could happen. I still can't believe it did happen. I believe it in my mind, but my heart still can't accept it."

Dale is buried among family in a small churchyard cemetery.

"Dale always said if he died that's where he wanted to be buried at," Janet said. "But before it had no meaning to talk about things like that. You don't ever think you're going to put that in action. Dale always told me brothers and sisters needed to be there for each other, that you can't take nothing for granted because you're never promised a tomorrow."

Doug Brown had a tomorrow, for which his wife, Meccah, was profoundly grateful, even as it made her hurt whenever she saw interviews or heard about a widow of one of Doug's less fortunate rig mates. And she knew Doug couldn't stop wondering why he made it out when others didn't. He couldn't stop feeling that, in some logic-defying way, his luck in surviving came at the expense of a friend's death.

Doug had survived, but the disaster that began on the Horizon was far from over for him. His future was engulfed by uncertainty. Eight months after his engine control room exploded, Doug still walked with a cane and struggled with extreme pain in his left knee as he considered undergoing yet another surgery, this one a knee replacement.

He was also dealing with the consequences of a post-concussion brain injury resulting from head trauma sustained in the blast. He needed special sunglasses for his eyes, which had become painfully light sensitive. His dog sat at his feet and whined because Doug could no longer take her for walks in the dog park like he used to. He didn't move the same way. His muscles twitched and his hands often grew shaky. Out for a drive, he sometimes forgot where he was headed. He had to write things down to be sure to remember them, and set alarms on his iPhone so he wouldn't forget to take his medications—Celebrex for the pain in his knee, something to help him sleep, and something else to help him wake up from the sleep medication.

Even with assistance, he rarely got a full night's rest. He was plagued by nightmares, consistent with his diagnosis of post-traumatic stress disorder. They were queasy, grasping dreams that could start anywhere, but always ended in the same dark place. He and Meccah and Kirah might be on vacation when suddenly the hotel where they were staying became the rig, which always ended up exploding as Doug tried to lead his family safely through the flames, only to get bogged down, as if in quicksand. After sleepless, dream-plagued nights, Doug would try to speak, and begin to stutter.

Doug had a team of professionals to assist him. He saw a counselor who monitored his program of medicines, a psychiatrist, a psychologist, and an orthopedist. It frustrated Doug that he couldn't undergo group therapy with other rig survivors, a kind of therapy that has proved particularly valuable for PTSD victims. The problem was twofold: All his former rig mates were scattered across the country, and they all had lawyers who didn't want them talking about what happened on the Horizon, however much it haunted them.

Doug had still been in a wheelchair in late May when he flew to Jackson, Mississippi, to participate in the Transocean-sponsored memorial service for the victims of the blowout. The company's chief executive officer, Steven Newman, told the gathered crew and their families, "This is the one of the most difficult days for many of us here. But for the families of our eleven lost colleagues, this is just another of many difficult days."

Country music star Trace Adkins, who had worked on an offshore rig as a young man, spoke via video link. "It was hard work, it was dangerous work," he said. "But nobody expects it to end like this."

A gospel choir sang hymns, an engraved ship's bell rang eleven times to mark each victim's death, and families were presented with one of the eleven bronze hard hats that ringed the stage.

In the space reserved for wheelchairs, Doug recognized Buddy Trahan, who had been critically injured and barely conscious the last time Doug had seen him, when they were both taken off the *Bankston* in a medevac helicopter. Doug wheeled over and bumped his chair against Buddy's.

"I don't know if you remember me," Doug began.

"Of course I remember you," Buddy said.

The solace Doug found in the company of his former rig mates that day has proved hard to hold on to.

"Not a day goes by that he doesn't have some kind of traumatic memory," Meccah said as 2010 came to a close. "He may get frustrated not being able to walk, or because he can't remember something, and he gets angry or sad. We've come to realize this is not something that's going to go away. We are just going to have to learn to live with it."

Doug's last day of full pay came in early December. Meccah helped him fill out the paperwork to start the disability pay. But as they understood the calculation, the disability benefits would deliver no more than 60 percent his salary. It was difficult to see how that would cover all their expenses.

Transocean and BP, though, were both recovering with surprising rapidity from the lows inflicted by investor reaction to the Horizon disaster. By the end of 2010, Transocean had recovered half the ground it had lost in the market, while BP had made up two-thirds of the decline in its stock price. But Doug would not share in that turn of fortune.

"At fifty-one, his career is gone," Meccah said. "It's hard to comprehend how our lives and Doug's life have changed so much. We don't know what the future is going to be like. We don't know how much care and therapy Doug's going to need."

Meccah had to think about finding a job, which was difficult for both practical and emotional reasons, as she had chosen to be a stay-at-home mom for her daughter, Kirah, and hadn't worked in years. Kirah was born as the Horizon was being built in Korea. She turned eleven in January. Doug has been in Kirah's life since her birth, but decided in October, exactly six months after the blowout, to legally adopt her.

It was a rare moment of joy in a period marked by deep frustration. When the Browns' freezer went on the fritz, Doug, a master mechanic who could field-strip refrigeration systems blindfolded, fumbled fruitlessly in his kitchen as he tried to place a simple bolt in the proper hole. He couldn't keep his hand steady.

Doug has filed a civil suit in the Horizon disaster, but as 2011 began, investigators were still picking over the rig's bones, and it was impossible to predict when, or if, the legal system will ever grant Doug any compensation for his loss.

"Everything is up in the air," Meccah said. "There's no planning in our life."

For former Horizon captain Curt Kuchta, life didn't feel so much suspended as it did preserved in amber. Even eight months after the Horizon burned, Curt was still at home, still spending hours working with his lawyer to answer questions from government agencies and preparing to testify in a phalanx of legal proceedings. His attorney had instructed him not to read or watch news about the rig explosion or its aftermath, or get involved in any way with the media. Even posting to Facebook was discouraged. But even on full pay and physically uninjured, Curt had found recovering from the blowout difficult, especially around the holidays. The silver lining had been his ability to spend time with his children. He'd even traded in his Grady-White fishing boat for a kid-friendly runabout.

From time to time he'd spoken with Transocean's marine manager, in charge of the nautical side of the company's rig operations. As 2011 dawned, Curt hoped to be returning to work; if not in the Gulf of Mexico, somewhere else would have to do.

Dave Young recovered from his boat racing mishap, a healing in some ways easier than getting over the nightmares and sleeplessness that had plagued him since his narrow escape from the burning rig. It helped to be able to spend time enjoying his family, and to finish building a new boat with some design wrinkles Alyssa knew he'd soon put to the test, however much she hoped he wouldn't.

But Dave needed the stimulation of his profession, and he had more than a few friends on Transocean rigs eager to have him back. Matt Michalski, Dave's former college classmate and captain of the Development Driller II, which had drilled one of the relief wells, requested that Dave be assigned to be his chief mate. Michalski says company managers told him they thought that might be rushing things. They wanted Dave to make a more gradual reentry to life offshore.

In the fall, Dave returned to work in Transocean's Houston headquarters. Living in a nearby hotel for three weeks at a stretch, as if he were on a rig, he clocked in nine to five, assisting the marine superintendents, shuffling paper, making calls, filling out forms, and commuting to the office like a white-collar worker.

Dave hated it but didn't complain. Instead, on his second hitch, he talked his way back into the Gulf, at least sporadically, choppering out to various rigs to audit maritime equipment and procedures—ensuring their safety—while he waited for a permanent assignment on a new rig.

Dave didn't know how he'd ever get over the loss of Jason and his other friends on the rig floor. He no longer was sure where he would end up himself. Not much about his professional future was crystal clear now, except maybe for one thing: Dave had no plans to sue anybody.

ACKNOWLEDGMENTS

In many ways this book was a group project and could never have been completed without the generous assistance of so many. We are especially indebted to those who had sailed on the Deepwater Horizon and their families for sharing often painful memories with us—Dave and Alyssa Young, Doug Brown, Curt Kuchta, Marcel Muise, Janet Woodson, Roger Burkeen, Rhonda Burkeen, Rebecca Wheeler, John Allen, and Huston Funk. The technological complexities of deepwater drilling in general and the Macondo well in particular presented a steep learning curve that could not have been scaled without the unstinting tutelage of Paul Parsons and Robert Almeida. Thanks, too, to oil field veterans Richard Robson, Matt Michalski, Peter Mello, and Alwin Landry for invaluable perspective. We were incredibly fortunate to have reporting assistance from two of the best in the business, Bill Rose and April Witt, and the keen reading eyes of Gene Weingarten and Lisa Shroder. We also benefited from the excellent reportage of the nation's newspapers, especially that of *The Washington Post*, *The New York Times*, and *The New Orleans Times-Picayune*, and Tom Junod of *Esquire* magazine.

Our gratitude goes to Gail Ross and Howard Yoon, for agenting, as well as hand-holding and tear-drying; and to our editors, David Hirshey and Barry Harbaugh, for guiding us through the shoals.

John Konrad would like add personal thanks to: his wife, Cindy, and Jack and Eleanor—for getting him through long nights writing and months spent offshore; Marcia, Jack, Maggie, Andrew, JD, Mairead, and his extended family for believing when writing a book seemed like the last thing he could accomplish; Mike Schuler and Tim Konrad for running gCaptain in his absence; Thad Fendley, Lee Freeman, Steve and Rachel Gordon, and the gCaptain.com forum members who answered so many of our questions; LCDR Tony Russell and ADM Thad Allen for providing U.S. Coast Guard access; Captain Dan Sheehan, for giving him the time to write; Richard DuMoulin, for his advice and support and introducing him to the wonders of life at sea; David, Lew, Steve, Dan, Eric, Wolf, Oscar, Ben, and countless other friends at Transocean and BP; the crews of the Spirit, Deep Seas, Ascension, and, especially, the D534, who welcomed him offshore and taught him all that he knows. And the crew of the *Bankston* and Coast Guard rescue teams who saved so many friends that day.

ABOUT THE AUTHORS

JOHN KONRAD is a veteran oil rig captain; a former employee of Transocean; and the founder of the world's leading maritime blog, gCaptain.com. A graduate of SUNY Maritime College, he lives in Morro Bay, California.

TOM SHRODER was an editor and writer at *The Washington Post* from 1999 to 2009. Under his stewardship, *The Washington Post Magazine* won the Pulitzer Prize for feature writing in both 2008 and 2010. He is the author of the nonfiction bestseller *Old Souls*, and lives in Vienna, Virginia.